Y0-AWG-366

Elementary Plane Geometry

Euclid (about 300 B.C.*) Euclid first collected the geometric facts known to the ancient Greeks and put them into an organized form in his work, the Elements. The Elements consists of 13 books and includes all of the important mathematics of early Greece. When asked by a student if geometry was practical, Euclid told his slave "Give him threepence, since he must make gain out of what he learns"* (Brown Brothers).

Elementary Plane Geometry

R. DAVID GUSTAFSON
PETER D. FRISK
Rock Valley College, Rockford, Illinois

JOHN WILEY & SONS

New York • Chichester • Brisbane • Toronto • Singapore

Library of Congress Cataloging in Publication Data:

Gustafson, Roy David, 1936-
Elementary plane geometry.

1. Geometry, Plane. I. Frisk, Peter D., 1942- joint author. II. Title.
QA455.G888 516′.22 72-5840
ISBN 0-471-33700-5

Printed in the United States of America
20 19 18 17 16

Dedicated to our Wives
Carol and Martha

And Our Children
Kristy, Steven, and Sarah

Preface

This book is intended primarily for the college student or for the individual who wishes a quick review of the fundamentals of high school geometry. The recent trend has been to make geometry textbooks more satisfying to the student who is equipped to appreciate the beauty found in mathematical rigor. Moreover, most high school geometry books assume that the student already knows many of the facts of geometry before he enters his first formal course, and that deductive proof is the only new concept to be presented.

However, this situation does not apply to the college where many students have no prior knowledge of geometry and have not shown previous interest or talent in mathematics. To these students, the instructor must present the subject material in an extremely short time—usually in a one-semester, three-hour course. Hence, it is necessary to get to the heart of the subject quickly and to condense the material so that it can be completed in about 45 lessons.

For these reasons this book is concise and basically traditional. Some mathematical elegance and rigor has been sacrificed to avoid the discussion becoming mired in preliminaries and detail. Many statements have been postulated that could have been proved. Certain concepts have not been emphasized, such as the distinction between equality and congruence, and the one between a line segment and its measure. We believe that such distinctions, if taken to extremes, often confuse the student more than they help him. If an instructor wishes to emphasize them, he can easily do so in his lectures.

The concise nature of the material permits the instructor to reteach concepts as necessary and still have ample time for testing.

Techniques of geometric construction are introduced early. Our experience shows that students enjoy this; it is fun and often challenging, and it shows a "practical" use of the subject material. From a mathematical standpoint, constructions answer many questions of existence.

The exercises are comprehensive but modest in number. Those needed for subsequent developments are marked with an asterisk.

This text has been used successfully at the two-year college level for a one semester, three-hour course.

We thank Janice Schlegel and Mrs. Betty Bell for their work in typing the manuscript, and William Hinrichs, Darrell Ropp, Josephine Shipley, Robert B. Eicken, David Hinde, and Kenneth Enochs for their helpful comments.

R. David Gustafson
Peter D. Frisk
Rockford, Illinois

Glossary of Symbols and Abbreviations

$\overline{AB}$	line segment AB
$\overrightarrow{AB}$	ray AB
$m(\overline{AB})$	measure of line segment AB
AB	line segment AB or the measure of line segment AB
$\angle 1$	angle 1 or measure of angle 1
$m(\angle 1)$	measure of angle 1
$=$	equal
$\neq$	not equal
$\perp$	perpendicular
bis	bisector
supp	supplementary
comp	complementary
$\cong$	congruent
$\triangle$	triangle
cpcte	corresponding parts of congruent triangles are equal
adj	adjacent
$\parallel$	parallel
$\nparallel$	not parallel
▱	parallelogram
$A(\quad)$	the area of
π	pi (3.14159 . . .)
$\sqrt{}$	the square root of
$\overset{\frown}{AB}$	arc AB
$<$	less than
$>$	greater than
$\rightarrow$	if . . . then . . .
$\therefore$	therefore
$\sim$	not or similar
m	slope of a nonvertical line
$^\circ$	degree(s)

Contents

Elementary Plane Geometry

1

Fundamental Concepts

The word "geometry" is from the Greek *geo*, meaning earth, and *metron*, meaning measure. The subject probably has its origins in early Egypt where earth measure was made necessary by the annual flooding of the Nile River and the need for yearly surveying. Euclid, a Greek who lived about 300 B.C., is considered to be the first individual who collected the isolated geometric facts of the day and put them in an organized form in his book, *The Elements*.

In this course in geometry, we will be studying more than just isolated geometric facts; we will be studying a system in which these facts are joined together in a logical, coherent way. Indeed, we will often be as concerned with the way in which these facts are related as we will be with the facts themselves. In geometry, as in all of mathematics, the thread that ties one idea with another is called *deductive reasoning*. Before we discuss deductive reasoning, however, we will discuss a different type of reasoning called *inductive reasoning*.

In the scientific laboratory an experiment is conducted and a certain outcome is observed. The experiment is then repeated and its outcome is again observed. After several repetitions and similar outcomes, the scientist will generalize his findings into what seems to be a true "If . . ., then . . ." statement. If I do this, then that will follow: if I heat water sufficiently, then it will boil. If I let go of a weight, then it will fall. This type of reasoning is called inductive reasoning.

Inductive reasoning is the foundation of experimental science, but it does suffer from a major drawback—it attempts to say something about *all* possible cases after examining only a *few* cases. For example, many people for many years have tried to find a cure for the common cold. They have come up with remedies ranging from wonder drugs to hot toddys. Yet each attempt has failed. On the basis of all these failures, would we be willing to conclude that no cure will ever be found? No, for there is always the possibility that the next attempt will succeed.

Inductive reasoning moves from the specific case to the general: every horse we've ever seen has four legs, therefore, *all* horses have four legs. Every car we've ever seen has four wheels, therefore, *all* cars have four wheels. But if we see an Isetta, we realize that reasoning inductively can be risky business.

In geometry we are more concerned with a different type of reasoning called *deductive reasoning*. We will talk about deductive reasoning more fully in Chapter 10 and will gain an understanding of it as we use it in the following chapters. But briefly we can say this: as opposed to inductive reasoning, deductive reasoning moves from the general to the specific. For example, if we know that *any* two numbers, added in either order, give the same sum, would we be willing to believe that $721+62$ is the same as $62+721$ without actually adding them? If we are assured of the spelling rule that the letter "Q" is always followed by the letter "U", would there be any doubt as to the second letter in the spelling of the word "quarter," or the fourth letter of the word "enquire?" Whenever we apply a general principle to a particular instance, we are using deductive reasoning.

If so much is to be based on these general statements, it seems reasonable that we consider the source of these statements. How are they determined?

In order for a statement to have meaning, each individual word or term that it contains must be understood. This cannot be accomplished by carefully defining each term used, for some words or terms must be left undefined. For example, if we did not know the meaning of the word "obstruction," we would go to a dictionary for help. In one dictionary, "obstruction" means "a barrier." If this word were strange as well, we would look it up. A "barrier," says our dictionary, is a "hindrance." A "hindrance" is an "impediment," an "impediment" is an "obstruction," and we are back where we started.

Since it is impossible to find an unending string of synonyms for any term, we will simply admit that some terms must, of necessity, be left undefined.

After we accept some words as fundamental undefined terms and then define other words in a more formal way, we accept as a basis for our discussion certain general properties about these terms. General statements that are accepted as a starting point are called *axioms* or *postulates*. Armed with these concepts—undefined terms, defined terms, axioms or postulates—we are ready to deduce some results, called *theorems*. To prove theorems we reason, using the rules of deductive logic, from a given hypothesis to a desired conclusion. The tools we can use are the undefined and defined terms, our axioms and postulates, and previously proven theorems.

In this course we accept the following words as undefined terms. Although we have a feeling for what they mean, we make no attempt to define them:

point, line, plane, set, between, closed, figure, intersection, curve, measure, side, interior, exterior, bisect, corresponding.

Using these undefined terms, we will now make some formal definitions.

Definition 1.1 A *triangle* is a closed three-sided figure.

Definition 1.2 The points of intersection of the sides of a triangle are called the *vertices* of the triangle.

Definition 1.3 An *equilateral triangle* is a triangle all of whose sides are of equal measure.

Definition 1.4 An *isosceles triangle* is a triangle that has at least two sides of equal measure. The third side is called the *base* of the triangle. The angle opposite the base is called the **vertex angle.**

Definition 1.5 A *scalene triangle* is a triangle that has no pair of sides with equal measure.

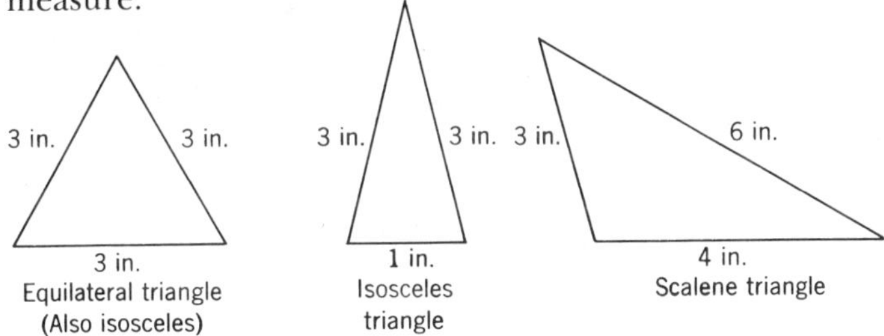

Equilateral triangle (Also isosceles)   Isosceles triangle   Scalene triangle

Definition 1.6 The *perimeter of a triangle* is the sum of the measures of the sides of a triangle.

Definition 1.7 Let A and B be two points on a line. The set of points on the line between A and B and including A and B is called the *line segment* AB. This segment is often denoted by $\overline{AB}$. The measure of $\overline{AB}$ is often written $m(\overline{AB})$.

Postulate 1.1 Two points determine one and only one line.

Definition 1.8 Let A and B be two points on a line. Point A together with all points on the B side of A is called the *ray* AB. We will denote this ray by $\overrightarrow{AB}$.

When we say that "$2+2=4$" we mean that "$2+2$" and "4" are just different names for exactly the same quantity.

Similarly, when we write "$\overline{AB} = \overline{CD}$", we mean that line segment AB and

Fig. 1.1 Thales (about 640–550 B.C.*). Thales was a very successful and widely traveled Greek merchant. In his travels, Thales learned the "practical geometry" known to the early Egyptians and helped bring the knowledge back to Greece. Thales has been called "the father of Greek mathematics, astronomy, and philosophy." One of his pupils was Pythagoras. (Radio Times Hulton Picture Library).*

line segment CD are exactly the same set of points. If we write $m(\overset{\bullet\!-\!\bullet}{AB}) = m(\overset{\bullet\!-\!\bullet}{CD})$, we mean the two line segments have equal numbers for their measures. When there is no chance for ambiguity, we will be careless and use the notation AB to stand for either the set of points in the line segment AB, i.e. $\overset{\bullet\!-\!\bullet}{AB}$, in the ray AB, i.e. $\overrightarrow{AB}$, in the line through A and B, i.e. $\overline{AB}$, or the measure of the line segment AB, i.e. $m(\overset{\bullet\!-\!\bullet}{AB})$, whichever is appropriate.

Notice that we have talked about the measure of $\overset{\bullet\!-\!\bullet}{AB}$, or $m(\overset{\bullet\!-\!\bullet}{AB})$, without talking about definite lengths. We are not interested here in whether the measure of AB is 3 inches, 2 millimeters, or 7 furlongs. We are interested in knowing when two line segments have the same measure, and we can often determine this without knowing what the measure is.

One of the problems of geometry is that of determining methods for drawing various geometric figures with certain properties. These methods are called geometric constructions, and often seem like puzzles or brain teasers because only two tools are allowed. One tool, called a straightedge, is used only for constructing lines. It differs from a ruler in that it may not be used to measure distances. The only time we are allowed to draw a line with a straightedge is when we know two points through which to draw the line. The other tool, a compass, has two uses. It may be used to transfer the measure of a line segment to another place on the paper, and it may be used to draw circles, or portions of circles, called *arcs*.

We will now consider procedures for copying a line segment and for finding the midpoint of a line segment. These constructions will use only a compass and a straightedge. Suppose that we have a given line segment, *AB*, but for some reason we would like to have a replica of segment *AB* located elsewhere on our paper. A very easy construction allows us to copy segment *AB* wherever we wish; such as on the line shown in Fig. 1.2 that passes through the point

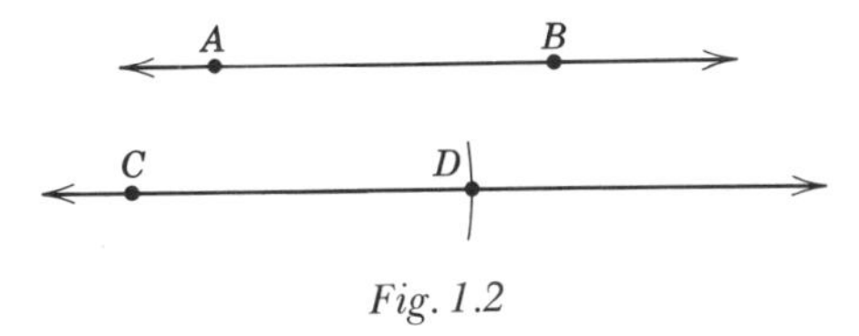

Fig. 1.2

C. We first put the point of our compass on point *A* and the pencil point of the compass on point *B*. After fastening the compass so that it is rigid, we transfer the measure of segment *AB* by placing the point of the compass at point *C* and drawing an arc with the pencil point. The intersection of this arc and the line determines a point *D* so that $m(\overline{AB}) = m(\overline{CD})$. Notice that technically $\overline{AB} \neq \overline{CD}$.

To bisect a line segment *AB*, set the compass to some convenient length (Fig. 1.3). This length must be greater than one-half the measure of the line

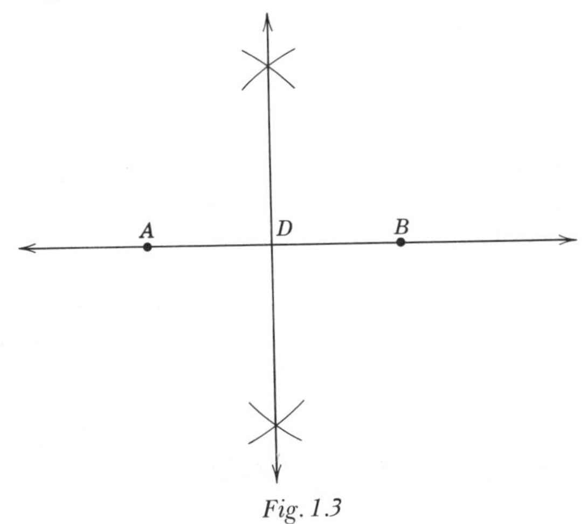

Fig. 1.3

segment AB. Placing the point of our compass at A, we make a large arc both above and below the line segment. Then placing our compass point at B, we again draw large arcs above and below the line segment. Draw a line through the two points where the arcs intersect. This line will divide segment AB into two line segments with equal measure: $m(\overline{AD}) = m(\overline{DB})$. We will later give justification for these steps and show that this line really does bisect $\overline{AB}$.

Definition 1.9 A line which divides a line segment into two segments with equal measure is called a *bisector of the line segment*. The intersection of the two is called the *midpoint* of the line segment.

By the preceding construction we see that any line segment has a midpoint. We will assume it has no more than one midpoint.

Postulate 1.2 A line segment has only one midpoint.

EXERCISES

1. Draw a triangle and construct a bisector on each of its sides. What do you discover about the bisectors? Would this be true for all possible bisectors?
2. Draw a line segment about 2 inches long. Divide the line segment into four equal parts.
3. What are the four elements of a deductive reasoning system?
4. Find the perimeter of a triangle with sides of 4, 5, and 6 inches.
5. Find the perimeter of an equilaterial triangle with a side of 15 inches.
6. The perimeter of an isosceles triangle is 32 inches and its base is 8 inches. Find the length of one of its equal sides.
7. The base of an isosceles triangle is half the length of one of the equal sides. The perimeter of the triangle is $89\frac{1}{2}$ inches. Find the length of the base.
8. Explain the distinction between the symbols $\overline{AB}$ and $m(\overline{AB})$.
9. Explain the distinction between a straightedge and a ruler.
10. Explain the distinction between inductive and deductive reasoning.
11. Explain the distinction between line, line segment, and ray.
12. Explain the necessity for undefined terms in a deductive reasoning system.
13. Explain the necessity for postulates in a deductive reasoning system.
14. Devise a procedure for constructing an equilateral triangle that has a given segment AB as a side.
15. Discuss the problem of creating a general law in the field of English grammar. For example, write a general law concerning the use of "ie" or

"ei" in proper spelling. Then see if your rule may be applied successfully to the words biennial, friend, receive, ceiling, weigh, neighbor, atheism, and seize. Why, in your opinion, is a general rule so difficult to formulate?

A very important figure in the study of geometry is the angle.

Definition 1.10 An *angle* is the union of two rays with a common endpoint. The common endpoint is called the *vertex* of the angle.

There are three different ways of naming an angle. The first way is to use letters to label three points on the angle—the vertex and a point on each ray. We would call Fig. 1.4 angle *QRP*, or angle *PRQ*. It is important to remember

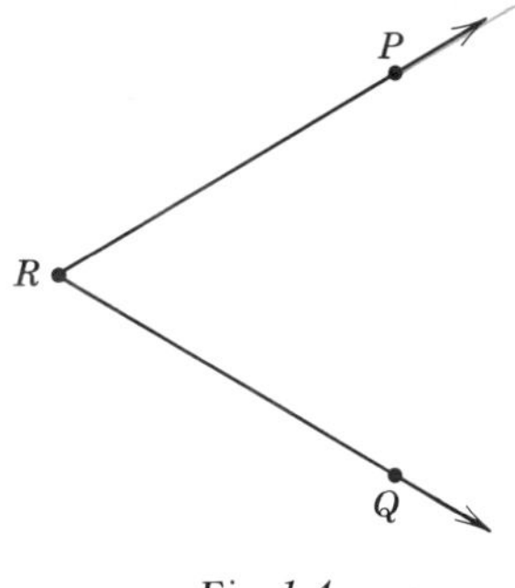

Fig. 1.4

that the name of the vertex point must always be between the names of the other points. From now on we will use the symbol "∠" to stand for the word "angle."

A second way to name an angle is simply to use the name of the vertex. In Fig. 1.4, $\angle PRQ$ could be named $\angle R$. Although this is a simpler way of naming angles, it can only be done if no ambiguity will arise.

A third way of naming an angle is to write a small number or a lower case letter in the interior of the angle. In Fig. 1.5 $\angle BCA$ can be named $\angle 1$. $\angle DEF$ in Fig. 1.6 can be named $\angle a$.

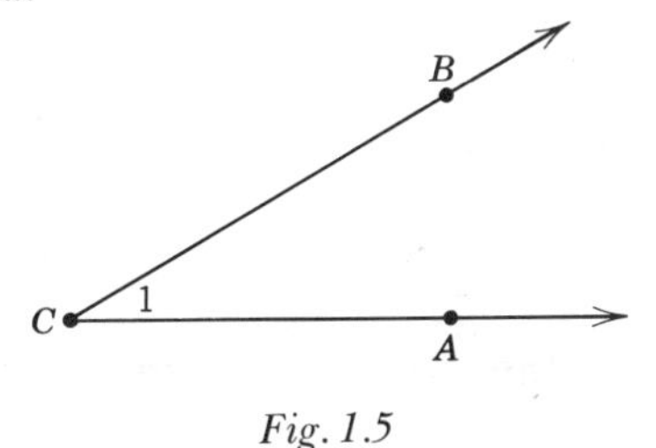

Fig. 1.5

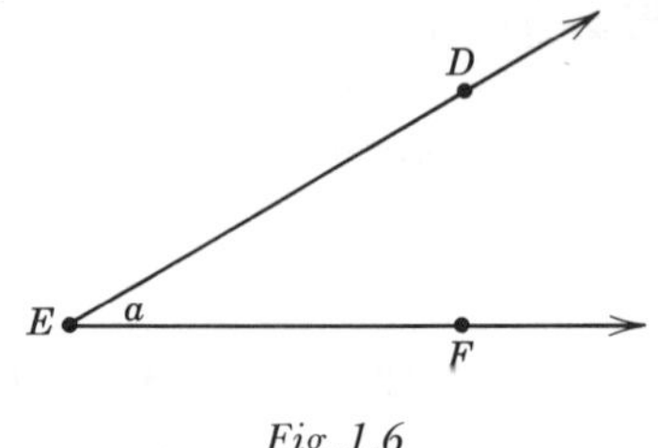

Fig. 1.6

Units of measure for line segments are commonly inches, feet, yards, meters, etc. One of the common units of measure for angles is the degree, denoted °. There are 360° in one complete rotation. One degree is 1/360th of a complete rotation. A 90° angle is $\frac{1}{4}$ of a complete rotation and a 180° angle is $\frac{1}{2}$ of a complete rotation.

Definition 1.11 A *straight angle* is an angle whose sides form a line. The measure of a straight angle is 180°.

Definition 1.12 A *right angle* is an angle whose measure is 90°.

Definition 1.13 An *acute angle* is an angle whose measure is larger than 0° but less than 90°.

Definition 1.14 An *obtuse angle* is an angle whose measure is greater than 90° but less than 180°.

As an example, there are several angles in Fig. 1.7.

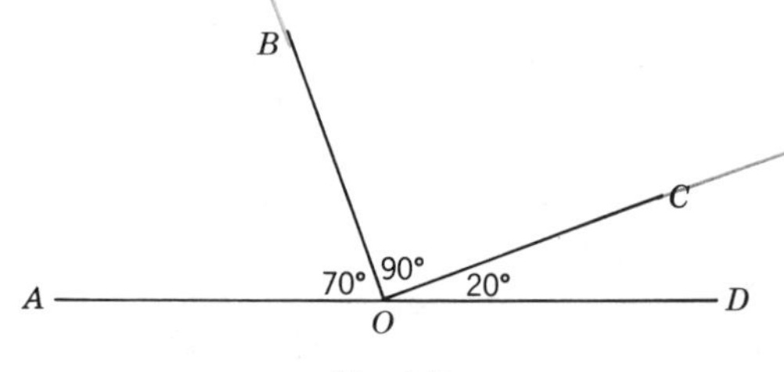

Fig. 1.7

$\angle AOB$ and $\angle COD$ are acute angles,
$\angle BOC$ is a right angle,
$\angle AOC$ and $\angle BOD$ are obtuse angles, and
$\angle AOD$ is a straight angle.

Many modern textbooks in geometry make a careful distinction between the word "equal" and another word, "congruent." As we have said, if two things are equal, they are identically the same. "$3+2=5$" means that "$3+2$"

and "5" are simply two different names for the same number. Similarly, "$\angle ABC = \angle DEF$" means that "$\angle$ABC" and "$\angle$DEF" are two different names for the same angle.

Two geometric figures are congruent if they are "identical twins," that is, if they have the same shape and the same measure. Since all line segments have the same shape and all angles have the same shape, the word congruent is often used to indicate that line segments or angles have equal measures. $\overleftrightarrow{AB}$ congruent to $\overleftrightarrow{CD}$ means $m(\overleftrightarrow{AB}) = m(\overleftrightarrow{CD})$. $\angle ABC$ congruent to $\angle DEF$ means $m(\angle ABC) = m(\angle DEF)$.

As we have already done with line segments, we will be careless and use the notation $\angle ABC$ to stand for both the set of points that forms the angle and the measure of the angle. Although we will make the distinction between equality and congruence from time to time to be certain that we remember, we will usually use the word "equal" to indicate that line segments or angles have equal measures.

The following axioms from algebra are useful:

Axiom 1.3 If equal quantities are added to equal quantities, the sums are equal quantities.

Axiom 1.4 If equal quantities are subtracted from equal quantities, the differences are equal quantities.

Axiom 1.5 If equal quantities are multiplied by equal quantities, the products are equal quantities.

Axiom 1.6 If equal quantities are divided by equal quantities, (not zero) the quotients are equal quantities.

Axiom 1.7 A quantity may be substituted for its equal in any mathematical expression.

Axiom 1.8 The whole is equal to the sum of its parts.

Axiom 1.9 If two quantities are equal to the same quantity, they are equal to each other.

Axiom 1.10 A quantity is equal to itself. (reflexive law)

Axiom 1.11 If a and b are any quantities, and $a = b$, then $b = a$. (symmetric law)

Theorem 1.1

All right angles are equal.

Proof: By definition each right angle has a measure of 90°. Therefore, all right angles have equal measure. Angles that have equal measure are congruent and therefore all right angles are congruent. However, as we have indicated, we will use the words "congruent" and "equal" as synonyms when discussing angles. Therefore, all right angles may be said to be equal.

Theorem 1.2

All straight angles are equal

The proof of this theorem is left as an exercise.

We have already discussed construction techniques for copying a line segment in another position and for bisecting a line segment. It is possible to devise similar techniques for angles.

To construct an angle with vertex at point B, equal to angle F, we set our compass at a convenient setting, place the point at F and describe an arc intersecting the sides of angle F at E and G (Fig. 1.8). Without changing the

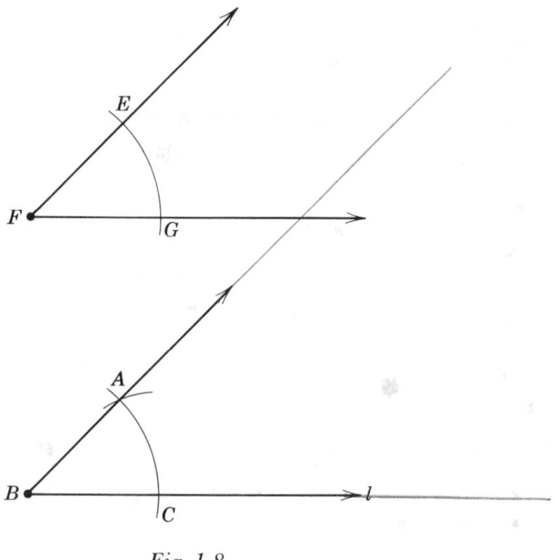

Fig. 1.8

compass setting, place its point at B, and mark a similar arc intersecting line l at point C. Using the compass now as a measuring device, set it so that the point is at G, and the pencil point is at E. Without changing this setting, place the point at C, and draw a small arc intersecting the larger arc. Call this point A. Draw ray BA. We will later prove that $m(\angle EFG) = m(\angle ABC)$.

Definition 1.15 An *angle bisector* is a ray that divides an angle into two angles that have equal measures.

There is a method for bisecting an angle using a compass and a straightedge that will be useful (Fig. 1.9). To divide $\angle ABC$ into two equal angles, place the point of the compass on point B and draw an arc intersecting the

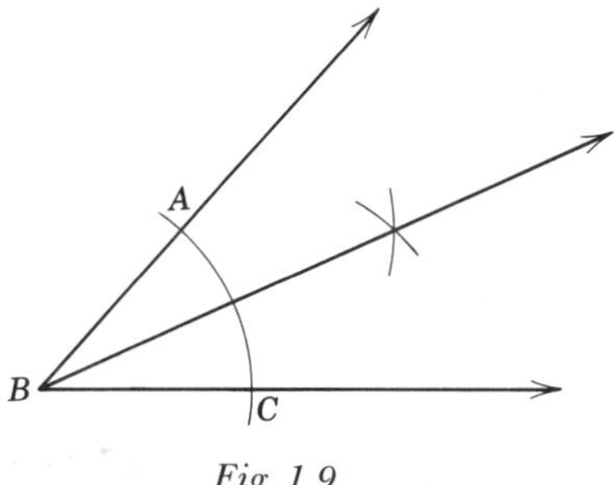

Fig. 1.9

sides of the angle at A and C. Now set the compass for some convenient length. Place the compass point on point A and draw a large arc in the interior of the angle. Now, without changing the setting of the compass, place the point on point C and draw a second large arc in the interior of the angle. Draw a ray joining the vertex of the angle to the point of intersection of the arcs. We will later prove that this ray is the bisector of the angle. Our construction demonstrates the existence of an angle bisector.

Postulate 1.12 The bisector of an angle is unique.

Two angles are said to be adjacent angles if they have a common vertex, and are side by side. In Fig. 1.10, $\angle 1$ and $\angle 2$ are adjacent angles.

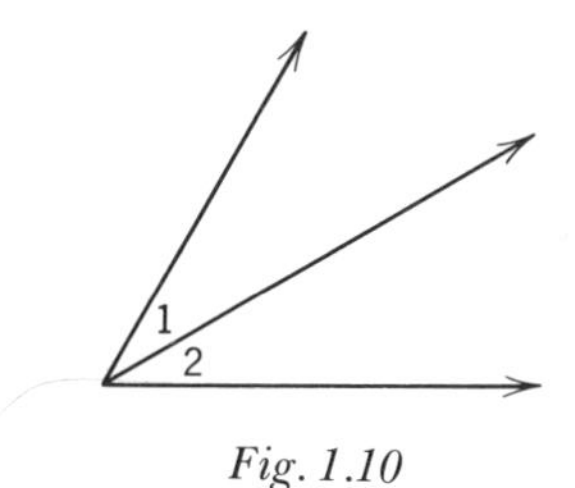

Fig. 1.10

Definition 1.16 Two angles are called *adjacent angles* if, and only if, they have a common vertex and a common side lying between them.

In Fig. 1.11, $\angle ABD$ and $\angle CBD$ are not adjacent angles. It is true that they have a common vertex and a common side, but the common side is not between the other sides. However, $\angle ABC$ and $\angle CBD$ are adjacent angles.

If two lines meet and form right angles, they are called perpendicular lines.

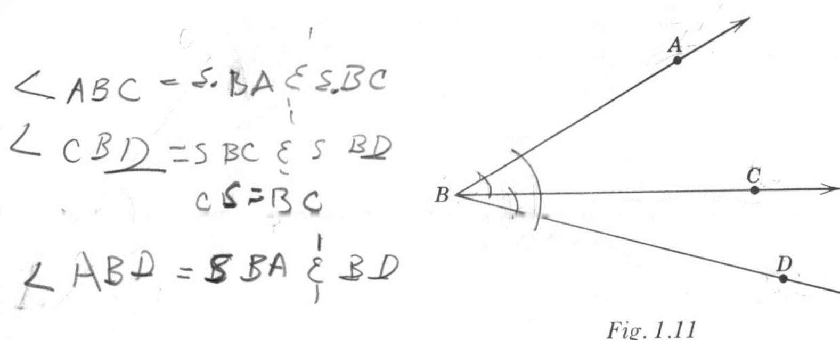

Fig. 1.11

A more useful wording of this statement is:

Definition 1.17 Two lines are *perpendicular* if, and only if, they meet and form equal adjacent angles.

In Fig. 1.12, the two lines are perpendicular if, and only if, $\angle 1 = \angle 2$. The symbol "$\perp$" means "is perpendicular to."

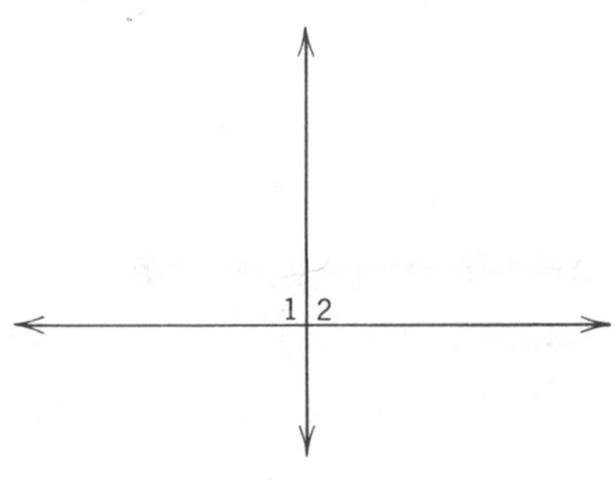

Fig. 1.12

Let A be a point on line l. To construct a perpendicular to line l through point A, we first set our compass at a convenient length. Place the point of the compass at A and make arcs intersecting line l at B and C. Now increase the setting of the compass. Placing the compass point at B and C draw large arcs above the line l. These arcs will intersect at some point, call it point P (Fig. 1.13). Draw line PA. We will prove later that PA is perpendicular to line l. This construction demonstrates that a perpendicular to a line does exist.

Definition 1.18 A *perpendicular bisector* of a line segment is a line that bisects and is perpendicular to the given line segment.

The construction technique to bisect a line segment given on page 5 will actually produce a perpendicular bisector of the line segment.

Postulate 1.13 The perpendicular bisector of a line segment is unique.

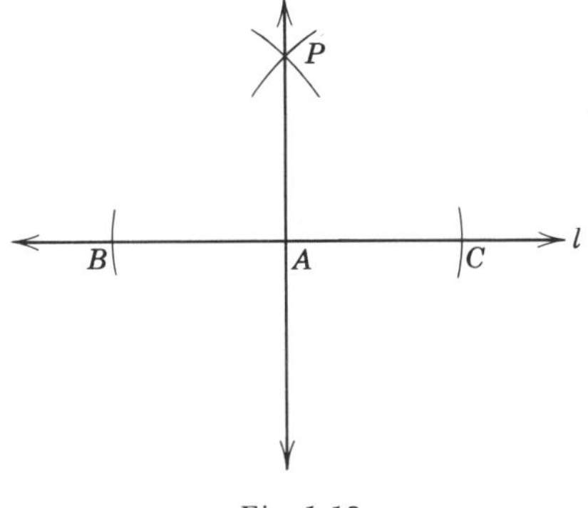

Fig. 1.13

Definition 1.19 A *right triangle* is a triangle with one right angle.
An *obtuse triangle* is a triangle with one obtuse angle.
An *acute triangle* is a triangle with all acute angles.

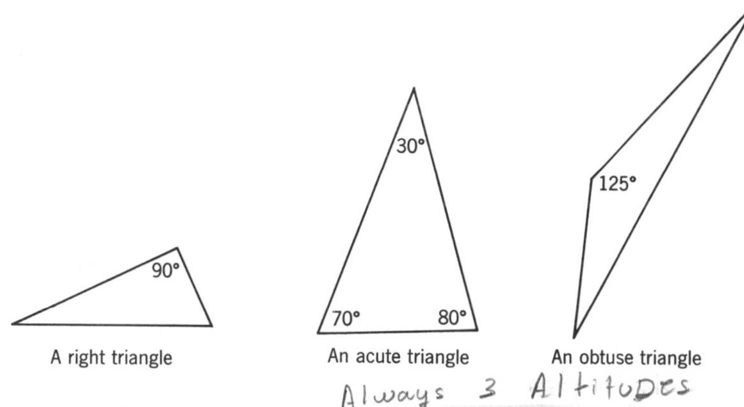

Definition 1.20 An *altitude of a triangle* is a line segment drawn from one vertex of the triangle perpendicular to the opposite side, or if necessary, to an extension of the opposite side. A *median of a triangle* is a line segment drawn from one vertex of a triangle to the midpoint of the opposite side.

As shown in Fig. 1.14, AP, BR, and CQ are altitudes of $\triangle ABC$. If M is the midpoint of side BC, then AM is one of three medians.

In order to construct an altitude of a triangle we must develop a procedure for the construction of a perpendicular to a line from a point not on the line.

Consider a line l and a point A not on line l (Fig. 1.15). Place the point of our compass on point A and draw an arc intersecting line l in two points. Call these points B and C. Place the point of our compass on point B and draw a large

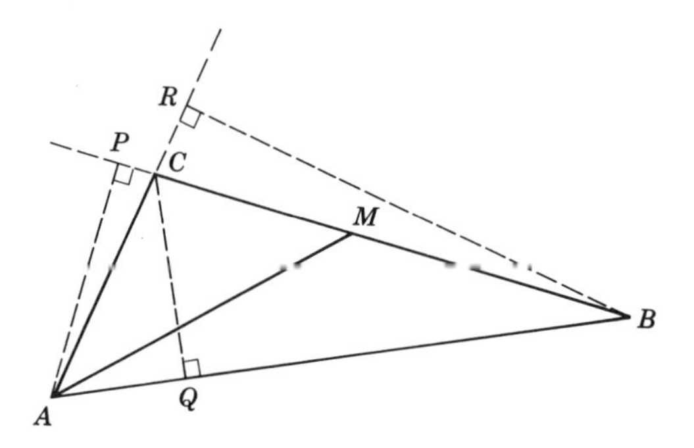

Fig. 1.14

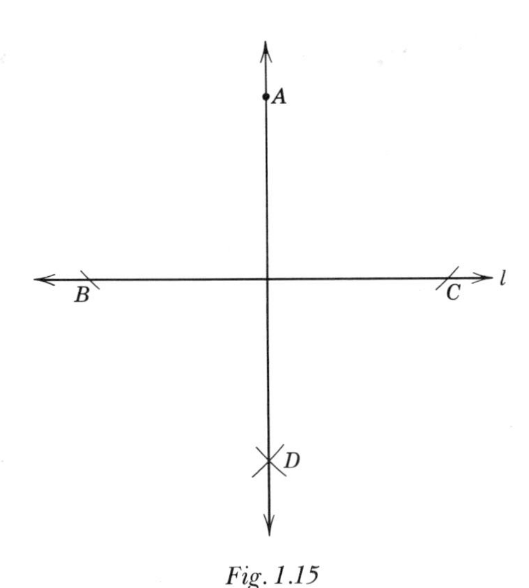

Fig. 1.15

arc below line l. Now place the point of the compass on point C and draw a second large arc below line l. Call the intersection of these arcs point D. Draw the line passing through points A and D. We will later prove that AD is the perpendicular to line l passing through point A.

Postulate 1.14 The perpendicular to a line through a point not on the line is unique.

EXERCISES

1. Is the size of an angle determined by the length of its sides?
2. Prove Theorem 1.2 on page 10.

3. Give another name for $\angle 2$.
4. Draw a triangle and bisect each of its angles. What do you discover about these bisectors?
5. Draw an angle and divide it into four equal angles.
6. List two pairs of adjacent angles.
7. Draw a triangle ABC. Construct the altitude on the base BC. Construct the other altitudes on the sides AC and AB. What do you discover about the three altitudes?
8. Draw a triangle ABC. Construct the three medians of the triangle. What do you discover about the medians?

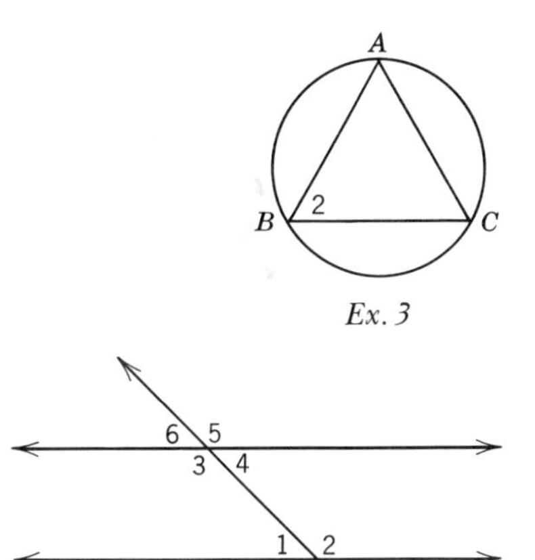

Ex. 3

Ex. 6

9. Draw a line l. At some point on line l construct a perpendicular, line t, to line l. At some point on line t construct a perpendicular, line r, to line t. What do you discover about line l and line r?
10. Explain the distinction between the symbols $\angle C$ and $m(\angle C)$.
11. Prove that two lines are perpendicular if, and only if, they meet and form right angles.
12. Draw examples of a right triangle, an obtuse triangle and an acute triangle.
13. A special case of Axiom 1.9 is the *transitive law*: if $a = b$, and $b = c$, then $a = c$. Use Axiom 1.9 to prove this.
14. *Prove:* the perpendicular to a line through a point on the line is unique. (*Hint:* use Postulate 1.13.)
15. Solve each of the following equations, justifying each step with one of the axioms on page 9.
    (a) $x - 2 = 4$
    (b) $x + 2 = 5$
    (c) $\frac{x}{3} = 2$
    (d) $2x = 6$
    (e) $2x - 3 = 7$
    (f) $\frac{x+3}{2} = 5$
16. If $x = y + 2$, and $y = 3$, what does x equal? What does $x + y$ equal? What axiom are you using to arrive at your answer?

Certain pairs of angles play very important roles in the study of geometry. Two such pairs are supplementary and complementary angles.

Definition 1.21 Two angles whose sum is $90°$ are called *complementary angles*. Each angle is called the complement of the other.
Two angles whose sum is $180°$ are called *supplementary angles*. Each angle is called the supplement of the other.

Some very important theorems can now be proved that will be extremely useful in future work.

Theorem 1.3

Supplements of the same or equal angles are themselves equal.

Proof: Part I

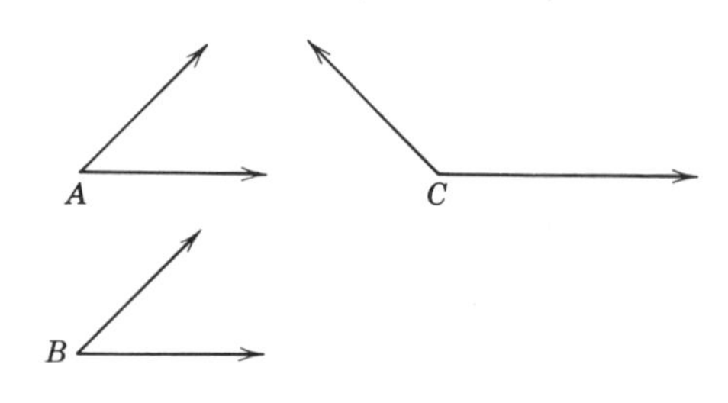

Let $\angle A$ be supplementary to $\angle C$ and let $\angle B$ be supplementary to $\angle C$. Then by the definition of supplementary angles, $\angle A + \angle C = 180°$ and $\angle B + \angle C = 180°$. By Axiom 1.9, $\angle A + \angle C = \angle B + \angle C$ and by Axiom 1.4, $\angle A = \angle B$.

Part II

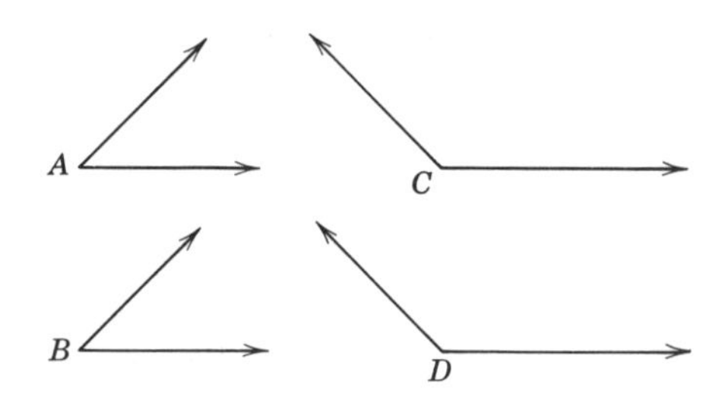

Let $\angle A$ be supplementary to $\angle C$ and let $\angle B$ be supplementary to $\angle D$ and let $\angle C = \angle D$. Then by the definition of supplementary angles, $\angle A + \angle C = 180°$ and $\angle B + \angle D = 180°$. By Axiom 1.9, $\angle A + \angle C = \angle B + \angle D$ and by Axiom 1.4, $\angle A = \angle B$.

Theorem 1.4

Complements of the same or equal angles are themselves equal.

Proof: Part I

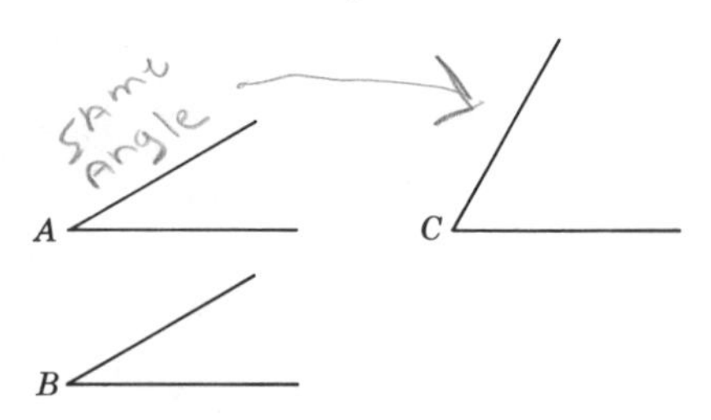

Let $\angle A$ be complementary and let $\angle B$ be complementary to $\angle C$. Then by definition of complementary angles, $\angle A + \angle C = 90°$ and $\angle B + \angle C = 90°$. By Axiom 1.9, $\angle A + \angle C = \angle B + \angle C$ and Axiom 1.4, $\angle A = \angle B$.

Fig. 1.16 Archimedes (287–212 B.C.*). Archimedes is famous for his work in geometry and applied mathematics. Many of his inventions were machines of war that were so successful that besieging armies were held off from Syracuse for three years. When the city was captured, the enemy soldiers of Claudius Marcellus were under orders not to kill Archimedes because of his scientific genius. However, a soldier found Archimedes drawing geometric figures in the sand and ordered him to move. When Archimedes ignored the order, he was killed. (Alinari–Art Reference Bureau).*

Part II

The case where complements of equal angles are equal is left as an exercise.

Definition 1.22 A pair of nonadjacent angles formed by two intersecting lines is called a *pair of vertical angles*.

In Fig. 1.17, $\angle 1$ and $\angle 2$ are a pair of vertical angles. Similarly $\angle 3$ and $\angle 4$ are a pair of vertical angles.

Fig. 1.17

Theorem 1.5

Vertical angles are equal.

Proof:

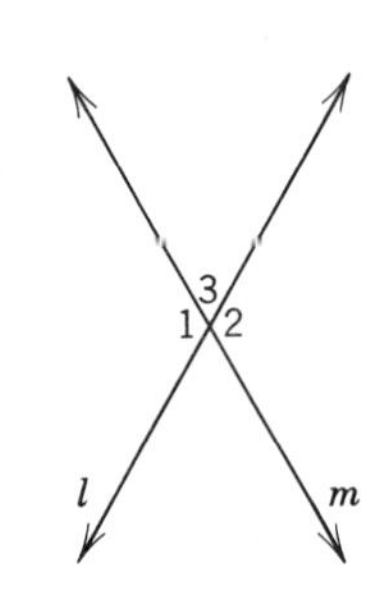

Let $\angle 1$ and $\angle 2$ be a pair of vertical angles. Since l and m are straight lines, $\angle 1 + \angle 3$ is a straight angle. Similarly $\angle 2 + \angle 3$ is a straight angle. Since the measure of a straight angle is 180°, $\angle 1 + \angle 3 = 180°$ and $\angle 2 + \angle 3 = 180°$. Therefore, $\angle 1$ is supplementary to $\angle 3$ and $\angle 2$ is also supplementary to $\angle 3$. By the theorem, supplements of the same or equal angles are themselves equal: $\angle 1 = \angle 2$.

EXERCISES

1. Prove Part II of Theorem 1.4: Complements of equal angles are themselves equal.
2. Find the supplement of an angle of (a) 30°; (b) 45°; (c) 135°.
3. Find the complement of an angle of (a) 30°; (b) 45°; (c) 80°.
4. Can an angle of 100° have a complement? A supplement? Why?
5. How large is $\angle 1$? $\angle 2$?
6. If two angles are equal, must they be vertical angles?
7. Must angles be adjacent to be complementary? supplementary?
8. Must angles be complementary or supplementary to be adjacent?
9. If $\angle 1 = \angle 2$, find the measure of $\angle 3$.
10. If $\angle ABD = 75°$ and $\angle CBD = 30°$, find the measure of the supplement of $\angle ABC$.

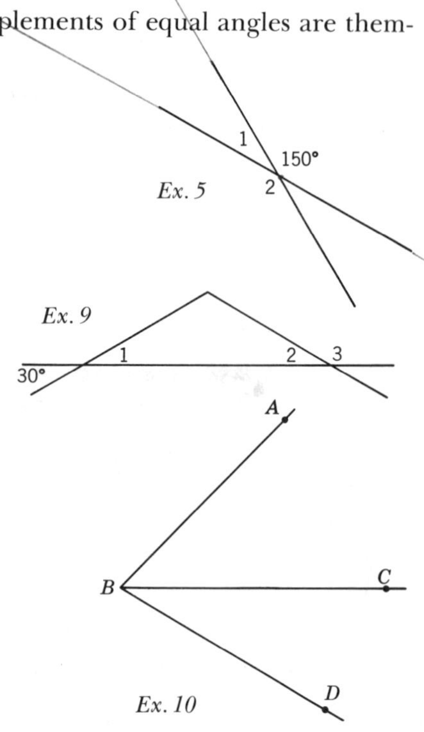

11. Find the complement of the supplement of an angle of 157°.
12. Find the complement of the complement of an angle of 37°.
*13. *Prove:* if two angles are supplementary and equal, then they are right angles.
14. If $\angle 1$ and $\angle 2$ are complementary, prove that $\angle 3$ and $\angle 4$ are complementary.
15. If AB bisects $\angle A$ and $\angle B$ and $\angle 1 = \angle 2$, what is the complement of $\angle ABC$?

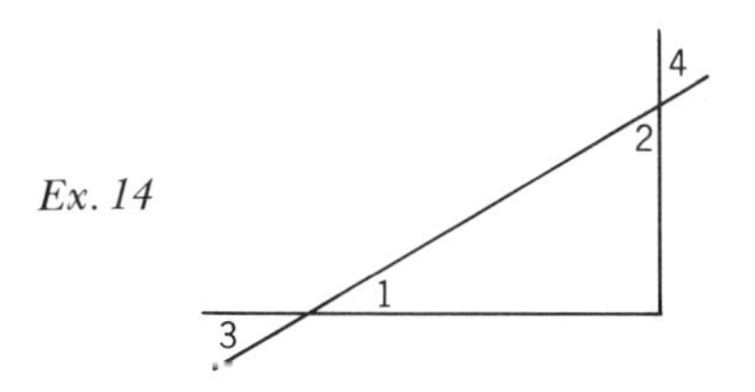

Ex. 14

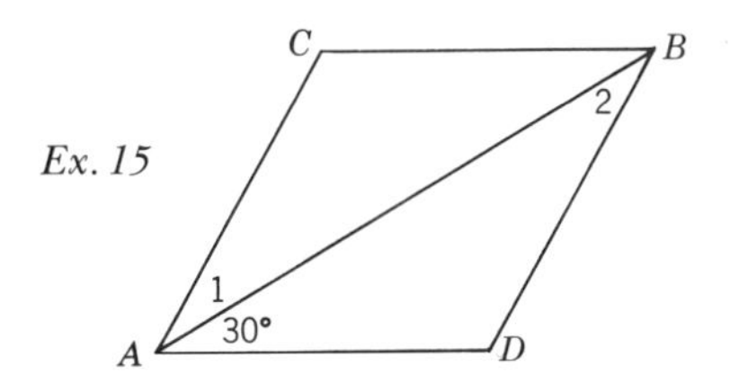

Ex. 15

Review Test

Classify the following statements as true or false. In this book a statement that is sometimes true and sometimes false is classified as a false statement.

1. Postulates are statements that have been proved.
2. The points of intersection of the sides of a triangle are called vertices of the triangle.
3. In a construction problem the only tools allowed are a compass and a straightedge.
4. A bisector of an angle divides the angle into equal parts.
5. An acute angle is greater than a right angle.
6. Angle ADB and $\angle ADC$ are a pair of adjacent angles.
7. Angle ADB and $\angle BDC$ are a pair of complementary angles.

Exs. 6–7

8. A line segment has only one bisector.
9. If two lines intersect, they must form two pairs of vertical angles.
10. An angle of 150° has no complement.
11. If two angles are complementary, they are also supplementary.
12. If two angles are complementary, their supplements are equal.

13. An acute triangle must have three acute angles.
14. If one line segment bisects a second line segment, the segments are perpendicular.
15. Adjacent angles may be supplementary.
16. $\angle ABC$ and $\angle ABD$ are complementary.
17. $\angle ABC$ and $\angle CBD$ are adjacent angles.

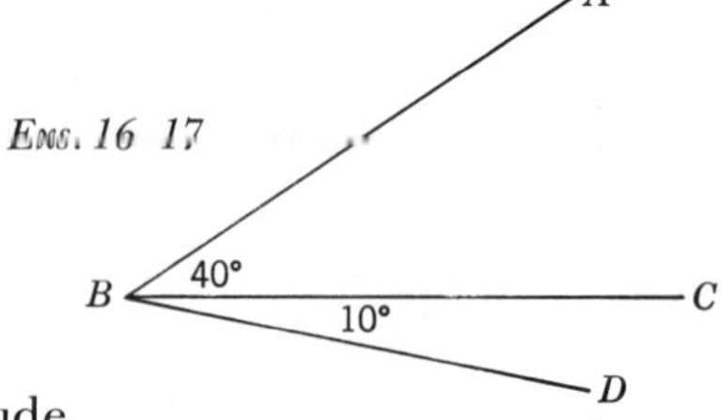

Exs. 16 17

18. An equilateral triangle has only one altitude.
19. An isosceles triangle may also be scalene.
20. A perpendicular to a line segment also bisects the line segment.
21. An isosceles triangle can be a right triangle.
22. Complements of complementary angles are equal.
23. An equilateral triangle is isosceles.
24. An angle has only one bisector.
25. The bisector of an obtuse angle divides the angle into two acute angles.

2

Congruent Triangles and Basic Theorems

In Chapter 1 we stated that two figures were congruent if they have the same size and the same shape. We then ignored the concept because all line segments have the same shape, all angles have the same shape, and only the measures are important. When we begin to talk about other figures, we will find some with the same size (area) but different shapes. We will find others with the same shape but different sizes (areas). We can no longer ignore the distinction between equal and congruent.

Definition 2.1 Two triangles are *congruent* if, and only if, all of their corresponding parts are equal. The symbol "≅" means "is congruent to."

If triangle ABC is congruent to triangle DEF, six equalities will follow:

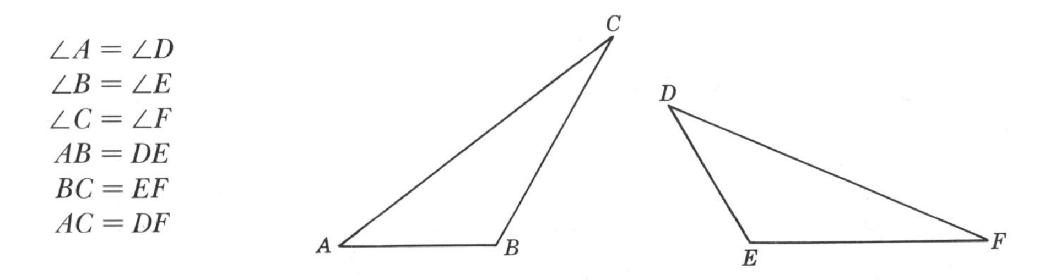

To prove that two triangles are congruent, we could prove that all six pairs of corresponding parts are equal. Not all of these six equalities are necessary; just three of the six are sufficient to prove two triangles congruent. It is very important, however, that we pick the three equalities in a very special way. Just any three equalities may be useless.

Postulate 2.1 If three sides of one triangle are equal, respectively, to three sides of a second triangle, the triangles are congruent. ($SSS = SSS$)

Postulate 2.2 If two sides and the included angle of one triangle are equal, respectively, to two sides and the included angle of a second triangle, the triangles are congruent. ($SAS = SAS$)

Postulate 2.3 If two angles and the included side of one triangle are equal, respectively, to two angles and the included side of a second triangle, the triangles are congruent. ($ASA = ASA$)

These three postulates are of extreme importance and should be memorized almost word for word. Students rarely have trouble with Postulate 2.1. However, they do have difficulty understanding what is meant by an included side or an included angle in Postulates 2.2 and 2.3. A side included between two angles is the segment joining the vertices of those two angles. An angle included between two sides is the angle that is formed by the two sides. For example, in $\triangle ABC$ (Fig. 2.1), side AB is included between $\angle A$ and $\angle B$. $\angle C$ is included between sides AC and BC.

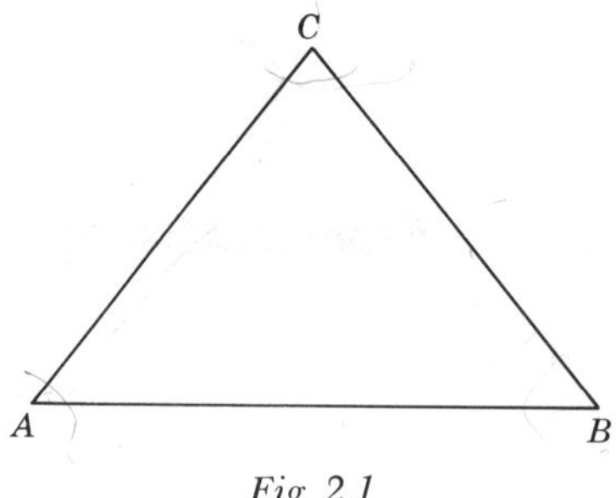

Fig. 2.1

We will now begin to do some practice problems in proving triangles congruent. In every case our plan will be to show that three equalities hold true and then to apply one of the above postulates. It will also be helpful if we follow a certain form in writing our proofs in these early stages. We will study the model proofs and pattern ours as closely as we can after them. In each case, we will be given some information. We may use this information, previous definitions, postulates, and theorems in our proofs.

In proving these exercises and future theorems, we must be careful not to assume too much about a geometric figure by looking at the drawing. For example, it is not correct to say that two line segments are equal just because they look equal. Also, if an angle looks like it is a right angle, we should not assume that it is a right angle.

Example 1

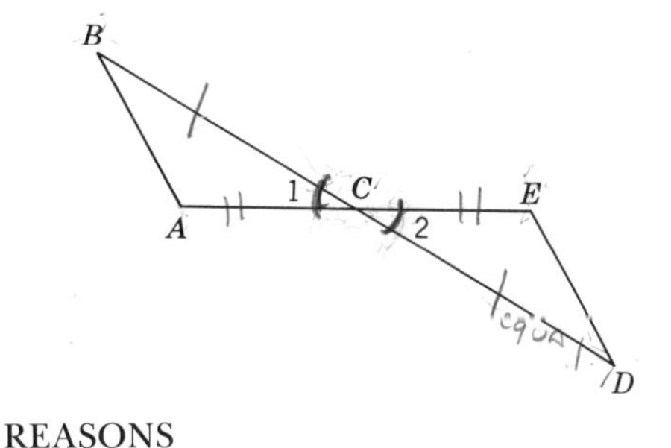

Given: AE and BD intersect at C
$BC = DC$
$AC = EC$

Prove: $\triangle ABC \cong \triangle EDC$

STATEMENTS	REASONS
1. AE and BD intersect at C.	1. Given.
2. $BC = DC$.	2. Given.
3. $AC = EC$.	3. Given.
4. $\angle 1$ and $\angle 2$ are vertical $\angle$s.	4. A pair of nonadjacent angles formed by two intersecting lines is called a pair of vertical angles.
5. $\angle 1 = \angle 2$.	5. Vertical angles are equal.
6. $\triangle ABC \cong \triangle EDC$.	6. If two sides and the included angle of one triangle are equal respectively to two sides and the included angle of a second triangle, the triangles are congruent.

Example 2

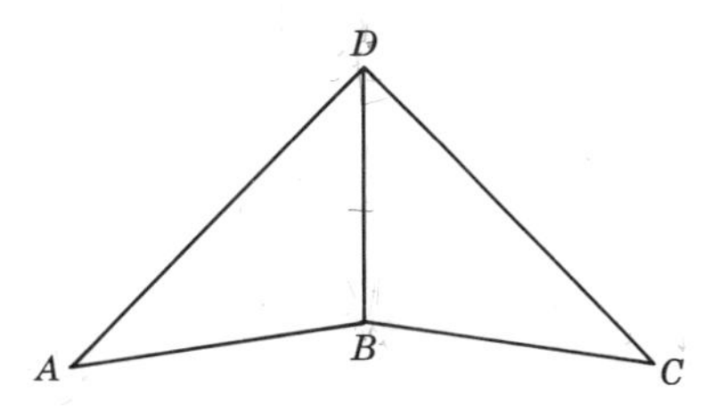

Given: $AD = CD$.
$AB = CB$.

Prove: $\triangle ABD \cong \triangle CBD$.

STATEMENTS	REASONS
1. $AD = CD$.	1. Given.
2. $AB = CB$.	2. Given.
3. $DB = DB$.	3. A quantity is equal to itself.
4. $\triangle ABD \cong \triangle CBD$.	4. If three sides of one triangle are equal, respectively, to three sides of a second triangle, the triangles are congruent.

Example 3

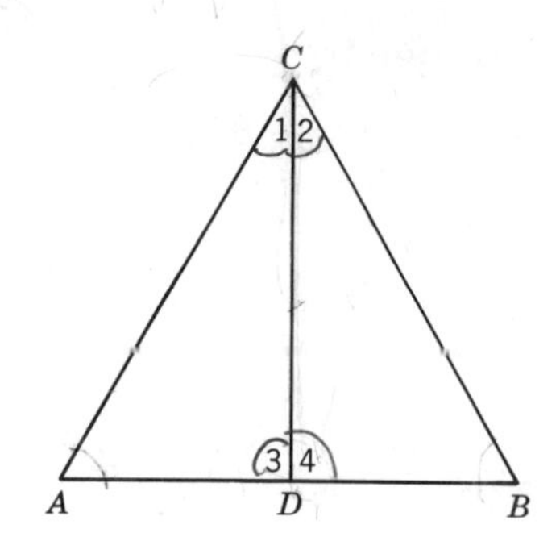

Given: $\angle 1 = \angle 2$.
$\angle 3 = \angle 4$.

Prove: $\triangle ADC \cong \triangle BDC$.

STATEMENTS	REASONS
1. $\angle 1 = \angle 2$.	1. Given.
2. $CD = CD$.	2. A quantity is equal to itself. Reflex
3. $\angle 3 = \angle 4$.	3. Given.
4. $\triangle ADC \cong \triangle BDC$.	4. If two angles and the included side of one triangle are equal, respectively, to two angles and the included side of a second triangle, the triangles are congruent.

Sometimes the triangles we wish to prove congruent are overlapping triangles. In this case it is sometimes helpful to use colored pencils to trace the triangles and keep things straight.

Example 4

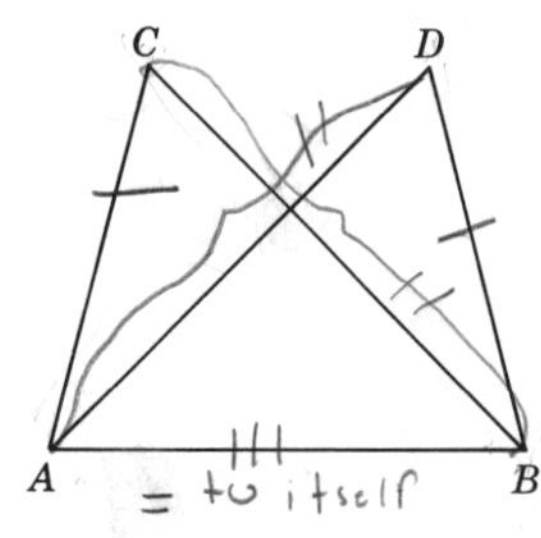

Given: $CA = DB$.
$CB = DA$.

Prove: $\triangle ABC \cong \triangle BAD$.

STATEMENTS	REASONS
1. $CA = DB$.	1. Given.
2. $CB = DA$.	2. Given.
3. $AB = BA$.	3. A quantity is equal to itself. Reflex
4. $\triangle ABC \cong \triangle BAD$.	4. If three sides of one triangle are equal respectively to the three sides of a second triangle, the triangles are congruent.

EXERCISES

Prove the following using the suggested form.

1. *Given:* $AD = BD$.
   $AC = BC$.

   Prove: $\triangle ADC \cong \triangle BDC$.

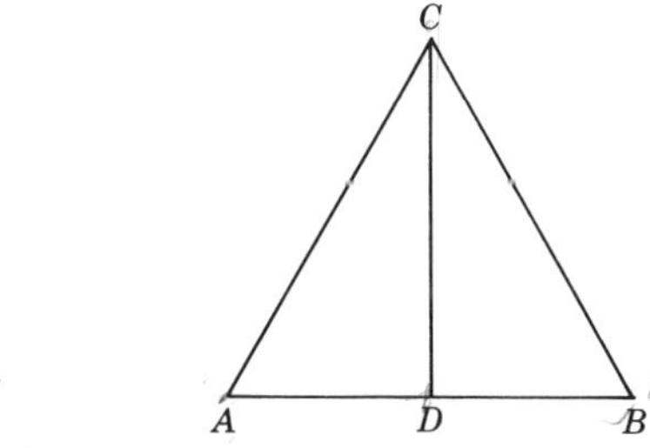

2. *Given:* AE and BD are straight lines intersecting at C.
   $BC = DC$.
   $\angle B = \angle D$.

   Prove: $\triangle ABC \cong \triangle EDC$.

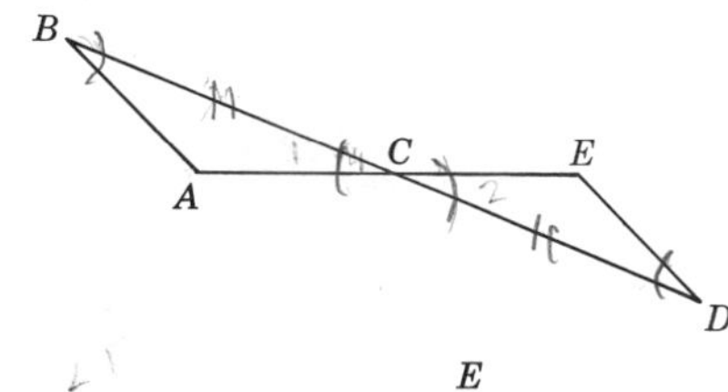

3. *Given:* $\angle A = \angle D$.
   $AE = DE$.
   $AC = DB$.

   Prove: $\triangle ABE \cong \triangle DCE$.
   (*Hint:* Consider Axiom 1.4.)

4. *Given:* $DC = BA$.
   $AD = CB$.

   Prove: $\triangle ABD \cong \triangle CDB$.

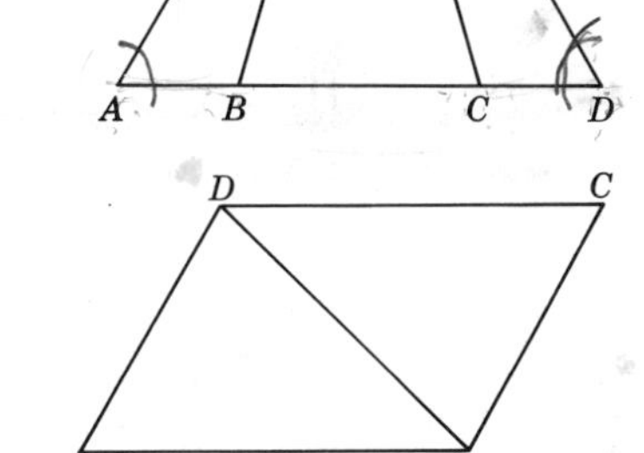

5. *Given:* $\triangle ADE$
   $BE = CE$.
   $\angle 1 = \angle 2$.
   $AB = DC$.

   Prove: $\triangle ABE \cong \triangle DCE$.

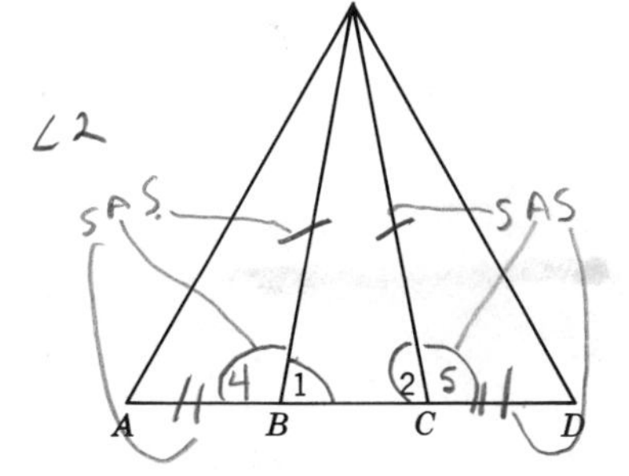

6. *Given:* $\angle A = \angle B$.
$AF = BE$.
$AC = BC$.

Prove: $\triangle AFC \cong \triangle BEC$.

7. *Given:* $CA = DB$.
$CB = DA$.

Prove: $\triangle ABC \cong \triangle BAD$.

8. *Given:* $AD = AC$.
$AB = AE$.

Prove: $\triangle ADB \cong \triangle ACE$.

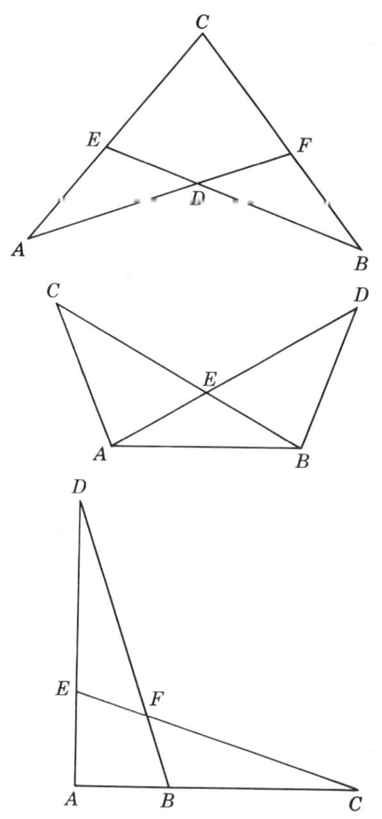

*9. Prove that two triangles that are congruent to a third triangle are congruent to each other.

It is not hard to extend this procedure to prove that corresponding parts of two triangles have the same measure. We would first prove the triangles congruent, and then make use of the definition of congruent triangles which states that all corresponding parts of congruent triangles are equal. A useful abbreviation for this statement is cpcte (corresponding parts of congruent triangles are equal).

Example 5

Given: $AC = BC$.
$\angle 1 = \angle 2$.

Prove: $\angle A = \angle B$.
Look for this

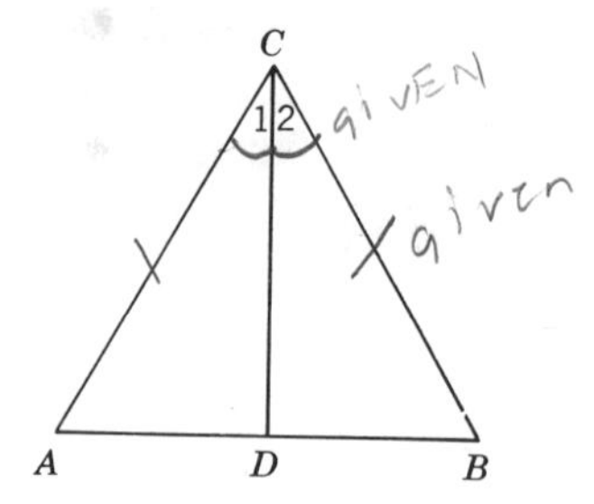

STATEMENTS	REASONS
1. $AC = BC$.	1. Given.
2. $\angle 1 = \angle 2$.	2. Given.
3. $CD = CD$.	3. Reflexive law.
4. $\triangle ACD \cong \triangle BCD$.	4. SAS = SAS.
5. $\angle A = \angle B$.	5. cpcte.

Example 6

Given: $AC = DB$.
$\angle 1 = \angle 2$.
$BE = CE$.

Prove: $\angle A = \angle D$

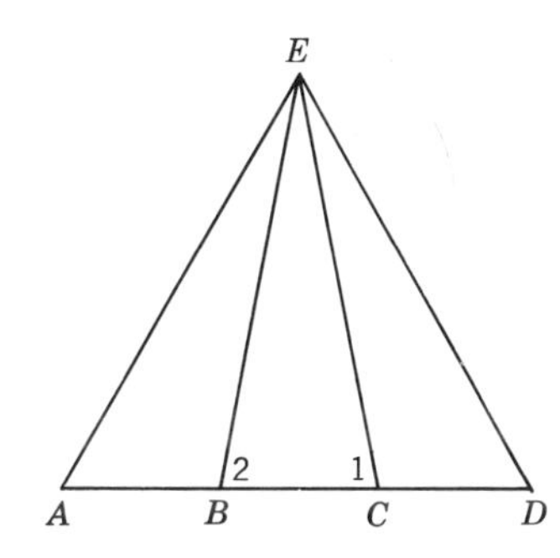

STATEMENTS	REASONS
1. $AC = DB$.	1. Given.
2. $BE = CE$.	2. Given.
3. $\angle 1 = \angle 2$.	3. Given.
4. $\triangle ACE \cong \triangle DBE$.	4. SAS = SAS.
5. $\angle A = \angle D$.	5. cpcte.

Example 7

Prove the construction of the bisector of a line segment given on page 5.

By construction: $AC = BC$
$AD = BD$.

Prove: CD bis AB.

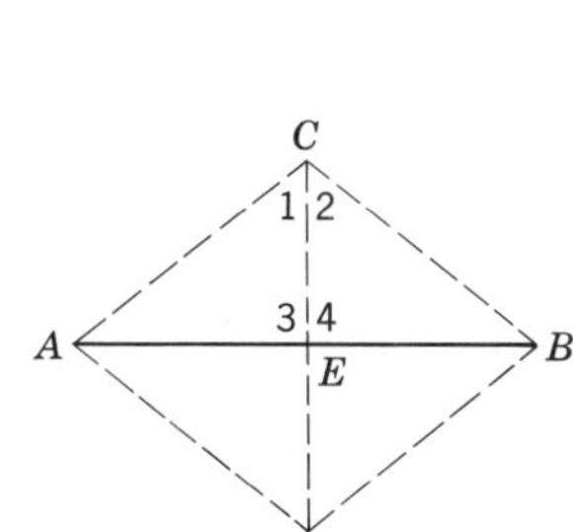
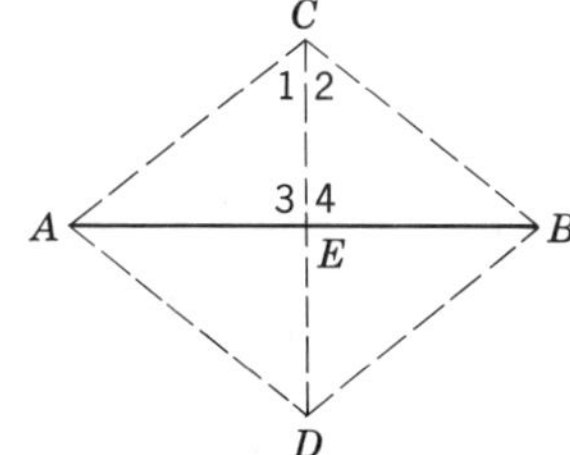

STATEMENTS	REASONS
1. $AC = BC$.	1. By construction
2. $AD = BD$.	2. By construction
3. $CD = CD$.	3. Why?

STATEMENTS	REASONS
4. $\triangle ADC \cong \triangle BDC$.	4. SSS = SSS.
5. $\angle 1 = \angle 2$.	5. Why?
6. $CE = CE$.	6. Why?
7. $\triangle ACE \cong \triangle BCE$.	7. SAS = SAS.
8. $AE = BE$.	8. Why?
9. CD bis AB.	9. Why?

EXERCISES

1. *Given:* AE and BD intersect at C.
$\angle A = \angle E$.
$AC = CE$.

Prove: $\angle B = \angle D$.

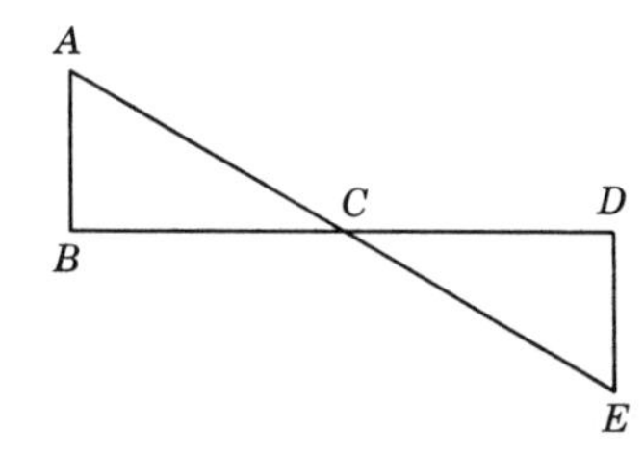

2. *Given:* $\triangle ADE$.
$\angle 1 = \angle 2$.
$AB = DC$.
$BE = CE$:

Prove: $AE = DE$.

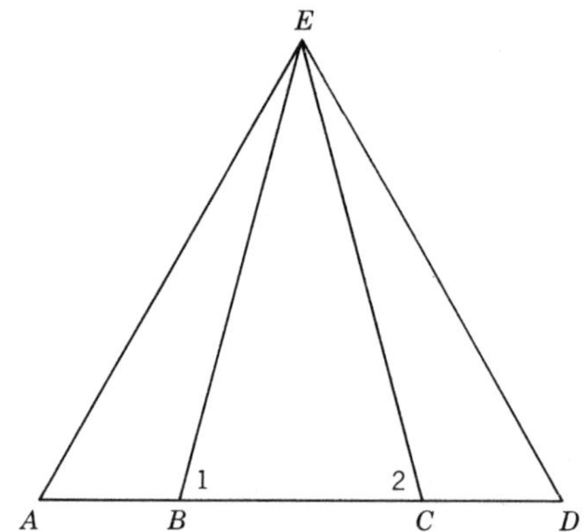

3. *Given:* $DB = EA$.
$AD = BE$.

Prove: $\angle DAB = \angle EBA$.

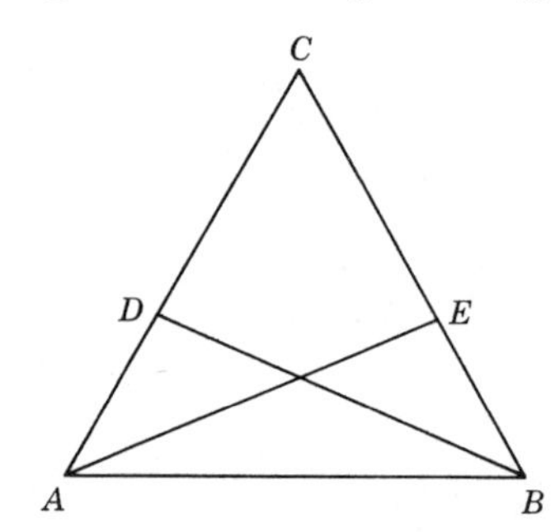

4. *Given:* $\triangle ABC$.
   $DC = EC$.
   $\angle 1 = \angle 2$.

   Prove: $AC = BC$.

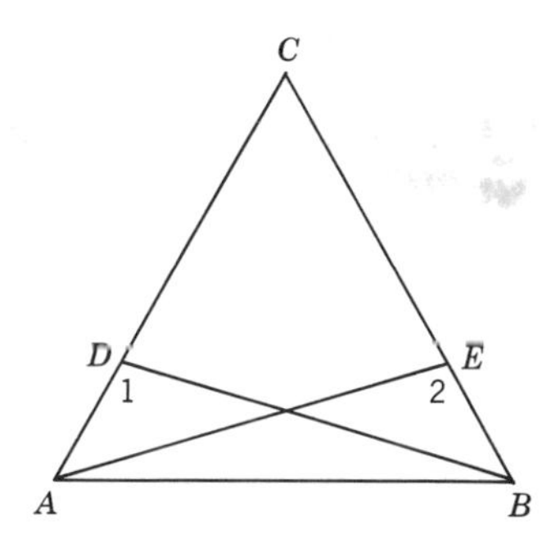

5. Prove the construction for copying an angle given on page 10.
6. Prove the construction for bisecting an angle given on page 11.
7. Prove the construction for a perpendicular to a line through a point on the line given on page 12.
8. Prove the construction for a perpendicular to a line through a point not on the line given on page 13.
9. *Given:* AC bisects DB at O.
   AC is perpendicular to BD.

   Prove: $DC = BC$.

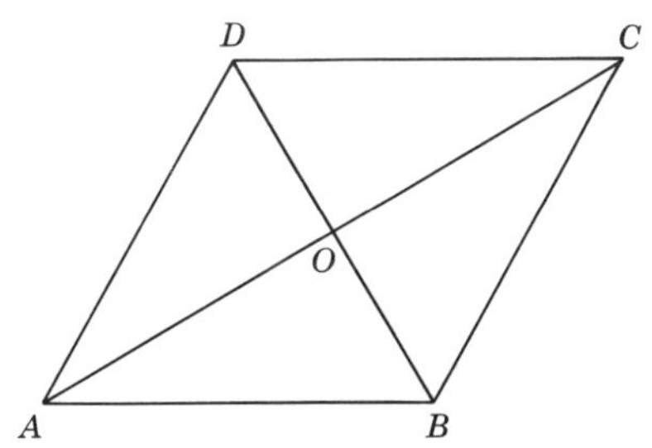

10. *Given:* $DC = AB$.
    $\angle CDA = \angle BAD$.

    Prove: $AC = BD$.

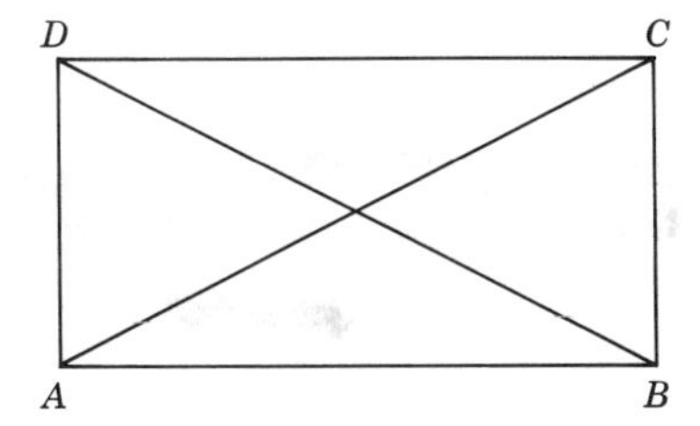

We are now ready to add extensively to our list of theorems. Our purpose is not to memorize the number associated with a theorem, but to understand what a theorem says. A good way to accomplish this is to state the theorems in our own words. It may be helpful to start a list of important theorems and definitions, adding to it as the book progresses.

Theorem 2.1

If two sides of a triangle are equal, then the angles opposite those sides are equal.

Given: $\triangle ABC$ with $AC = BC$.

Prove: $\angle A = \angle B$.

Construction. Construct the bisector of angle C.

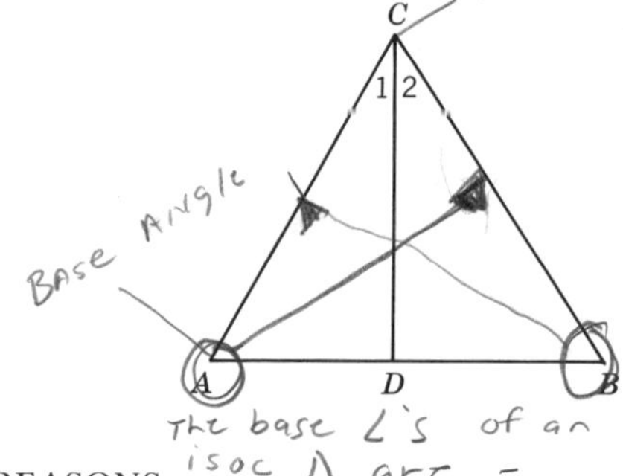

Proof:

STATEMENTS	REASONS
1. $\triangle ABC$ with $AC = BC$.	1. Given.
2. CD bis $\angle C$.	2. By construction.
3. $\angle 1 = \angle 2$.	3. An angle bisector divides an angle into two equal angles.
4. $CD = CD$.	4. Reflexive law.
5. $\triangle ACD \cong \triangle BCD$.	5. SAS = SAS.
6. $\angle A = \angle B$.	6. cpcte.

Notice that the triangle of Theorem 2.1 is isosceles. The equal angles of an isosceles triangle are often called base angles. This theorem may be stated more concisely as, "The base angles of an isosceles triangle are equal."

Theorem 2.2

The bisector of the vertex angle of an isosceles triangle is the perpendicular bisector of the base of the triangle.

Given: $\triangle ABC$ is isosceles with $AC = BC$; CD bisects $\angle ACB$.

Prove: CD is perpendicular bisector of AB.

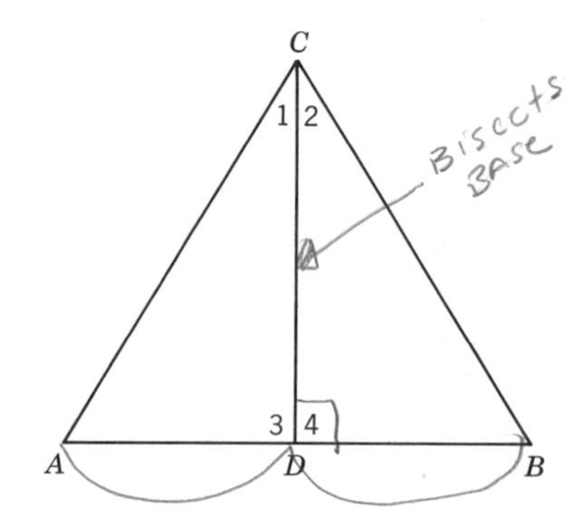

Proof:

STATEMENTS	REASONS
1. ABC is an isosceles triangle with $AC = BC$.	1. Given.
2. CD bisects $\angle ACB$.	2. Given.
3. $\angle 1 = \angle 2$.	3. An angle bisector divides an angle into two equal parts.
4. $\angle A = \angle B$.	4. The base angles of an isosceles triangle are equal.
5. $\triangle ACD \cong \triangle BCD$.	5. ASA = ASA.
6. $\angle 3 = \angle 4$.	6. cpcte.
7. $\angle 3$ adj $\angle 4$.	7. Def. of adjacent angles.
8. $CD \perp AB$.	8. Why?
9. $AD = BD$.	9. cpcte.
10. CD bis AB.	10. Why?
11. CD is perpendicular bisector of AB.	11. Steps 8 and 10.

Theorem 2.3 = omit

The perpendicular bisector of the base of an isosceles triangle passes through the vertex.

Given: Isosceles $\triangle ABC$ with base AB; DF is the perpendicular bisector of AB.

Prove: DF passes through point C.

Construction. Construct the angle bisector of the vertex angle C.

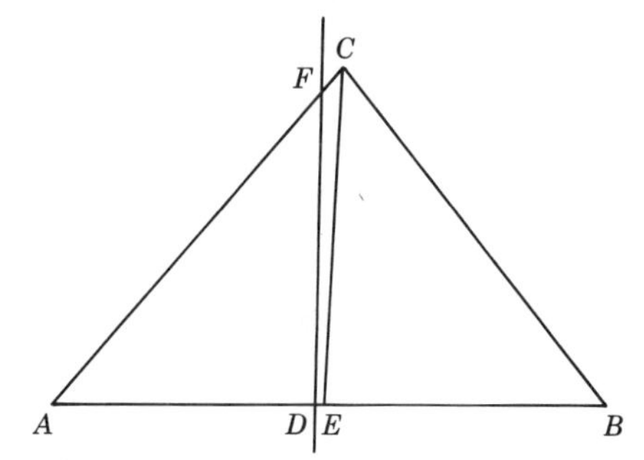

Proof:

STATEMENTS	REASONS
1. ABC is an isosceles triangle with base AB.	1. Given.
2. CE is bisector of $\angle ACB$.	2. Possible construction.

STATEMENTS	REASONS
3. CE is perpendicular bisector of AB.	3. Why?
4. DF is perpendicular bisector of AB.	4. Given.
5. CE and DF are the same line.	5. The perpendicular bisector of a line segment is unique.
6. DF passes through the vertex C.	6. Why?

Since two lines were shown to coincide, this type of proof is called *proof by coincidence*.

EXERCISES

1. *Prove:* an equilateral triangle has three equal angles. (A triangle with three equal angles is called an *equiangular* triangle.)
2. *Prove:* the median drawn to the base of an isosceles triangle is perpendicular to the base.
3. *Prove:* the median drawn to the base of an isosceles triangle bisects the vertex angle.
4. *Given:* ABC is an isosceles triangle with base AB. D is midpoint of AC. E is midpoint of BC.

   Prove: $\angle 1 = \angle 2$.

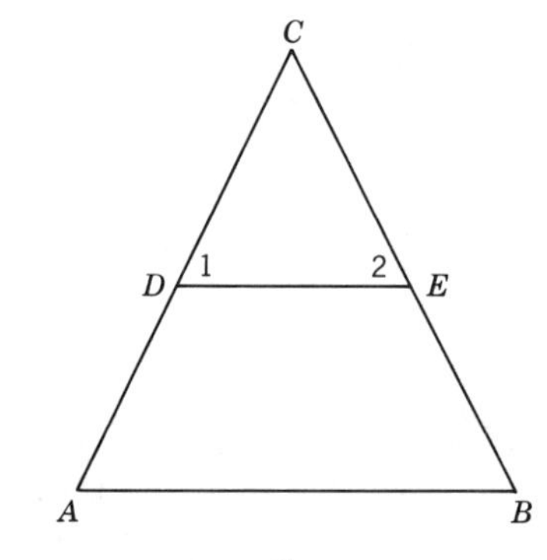

5. *Given:* $\angle 1 = \angle 2$.
   $AC = BC$.

   Prove: $\angle 3 = \angle 4$.

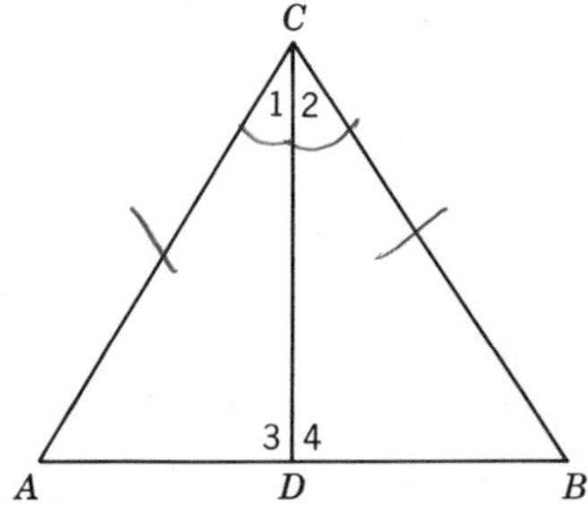

6. *Given:* $AC = BC$.
   $AD = BD$.

   Prove: $\angle CAD = \angle CBD$.

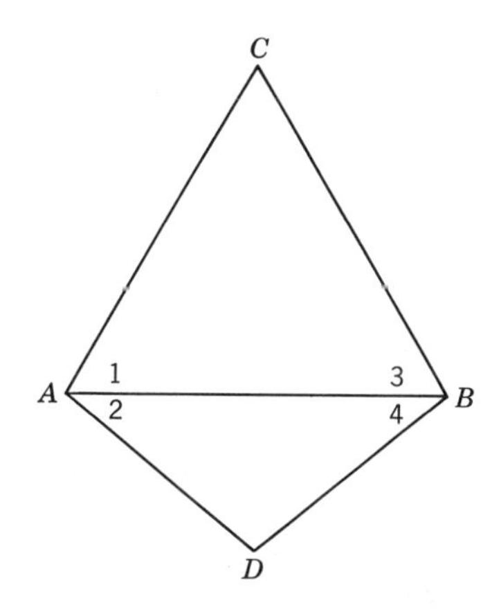

7. *Given:* $\triangle ADE$.
   $AE = DE$.
   $AB = DC$.

   Prove: $\angle 1 = \angle 2$.

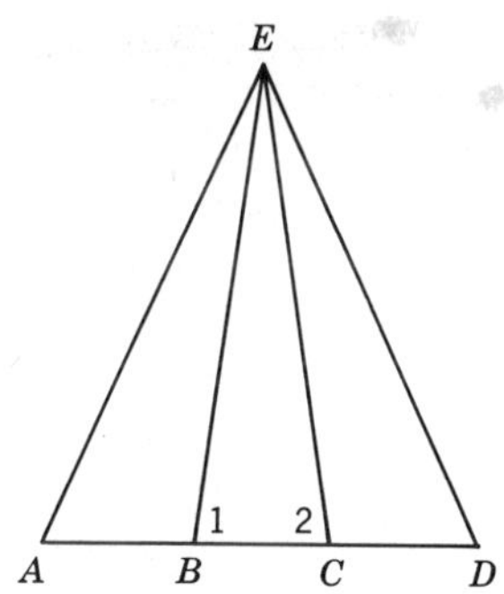

8. *Given:* Isosceles $\triangle ABC$, with $AC = BC$;
   CE bisects AB.

   Prove: $\triangle ABD$ is isosceles.

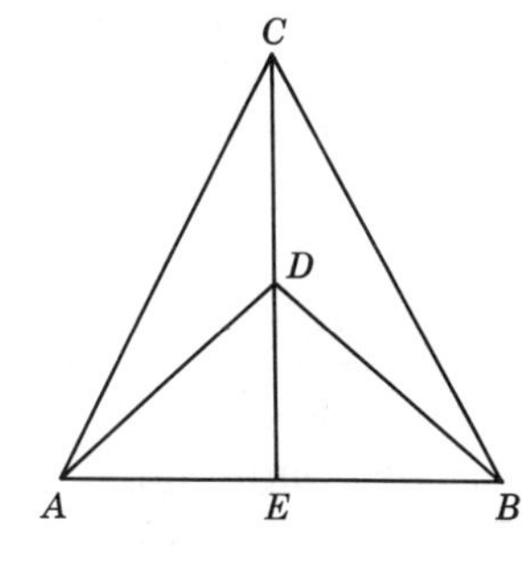

9. *Given:* Equilateral triangle ABC.
   D is midpoint of AB.
   E is midpoint of BC.
   F is midpoint of CA.

   Prove: $\triangle ADF$, $\triangle BED$, $\triangle$CFE, are all congruent, and that $\triangle DEF$ is equilateral.

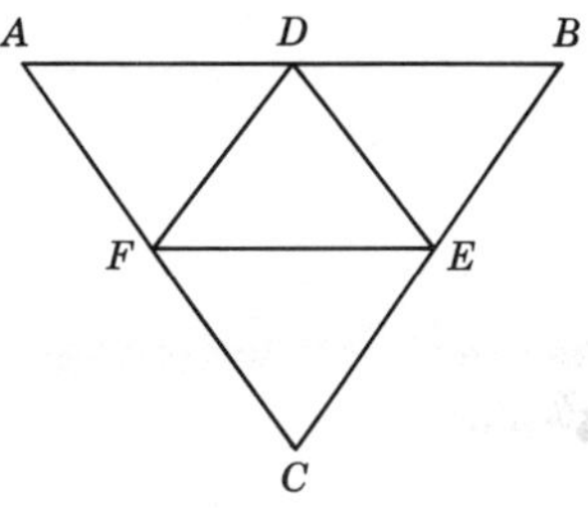

10. *Given:* Isosceles triangles ABC and ABD, with common base AB.

Prove: $\angle 1 = \angle 2$.

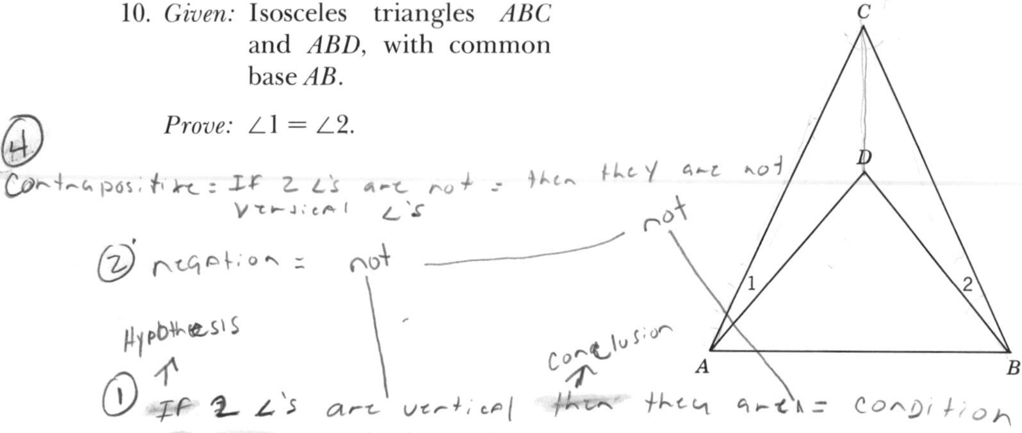

Any statement in the form of an "If . . . then . . ." statement is called a *conditional* statement. In Theorem 2.1 we proved the conditional statement: if a triangle has two equal sides, then the angles opposite those sides are equal. If the hypothesis (clause following the word "if") and the conclusion (clause following the word "then") are interchanged, a new statement is formed that is called the *converse* of the original statement.

The converse of Theorem 2.1 is: if a triangle has two equal angles, then the sides opposite those angles are equal.

If a conditional statement is true, its converse statement is not necessarily true. For example: "if two angles are vertical angles, then they must be equal" is a true statement. The converse statement, "if two angles are equal, then they must be vertical angles," is not necessarily true.

Also, "if Pat Smith is a woman, then Pat Smith is a human being," is a true statement. The converse, "if Pat Smith is a human being, then Pat Smith is a woman", may or may not be true. Since the converse does not follow from the conditional, we must prove the converse of a theorem just as we had to prove the original theorem.

In many of the exercises and theorems encountered thus far, we have shown certain triangles to be congruent. We have always worked with two triangles, although the two often shared a common side, or even overlapped. There is no reason, however, why two distinct triangles need to be present; it is sometimes useful to prove that a triangle is congruent to itself. This is done in the next theorem, which is the converse of Theorem 2.1.

Theorem 2.4

If a triangle has two equal angles, then the sides opposite those angles are equal.

Given: $\triangle ABC$ with $\angle A = \angle B$

Prove: $BC = AC$

Plan. We shall prove $\triangle ABC$ congruent to itself in such a way that side AC corresponds to side BC.

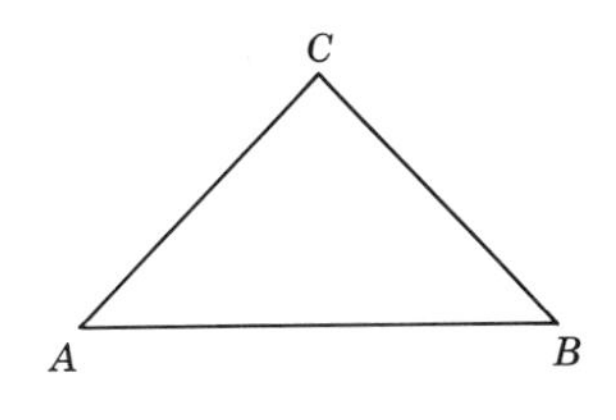

Proof:

STATEMENTS	REASONS
1. $\angle A = \angle B$.	1. Given.
2. $\angle B = \angle A$.	2. Symmetric property of equality.
3. $AB = BA$.	3. Reflexive property of equality.
4. $\triangle ACB \cong \triangle BCA$.	4. ASA = ASA.
5. $AC = BC$.	5. cpcte.

EXERCISES

1. *Given:* $\triangle ABE$.
   $AE = BE$.
   $AC = BD$.

   Prove: $\angle 3 = \angle 4$.

2. *Given:* $\triangle ABE$.
   $\angle 1 = \angle 2$.
   $AC = BD$.

   Prove: $\angle A = \angle B$.

3. *Given:* $\angle A = \angle B$.
   $AD = BE$.

   Prove: $\angle CDE = \angle CED$.

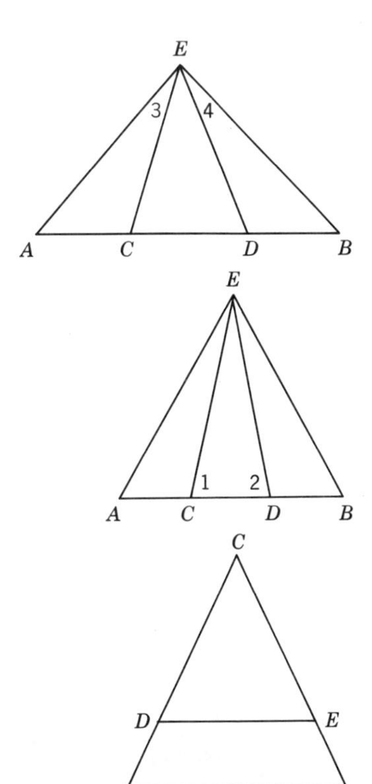

4. *Given:* $\angle 1 = \angle 2$.
$\angle 3 = \angle 4$.

Prove: $\angle CAD = \angle CBD$.

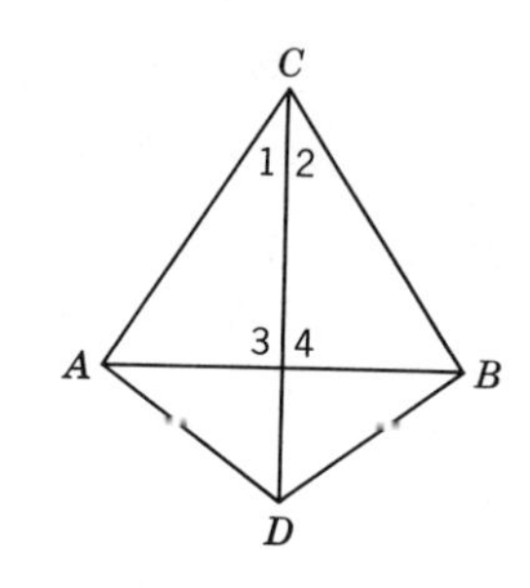

5. *Given:* DA bisects $\angle CAB$.
DB bisects $\angle CBA$.
$\angle 1 = \angle 2$.

Prove: $CA = CB$.

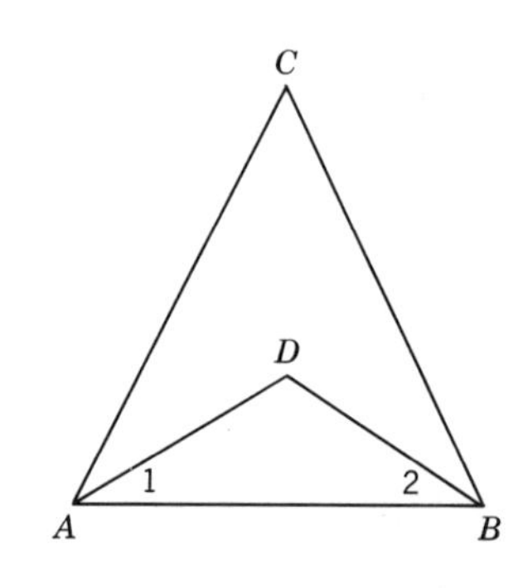

6. *Given:* $\angle 1 = \angle 2$.
$\angle 3 = \angle 4$.

Prove: $\triangle ADC \cong \triangle BEC$.

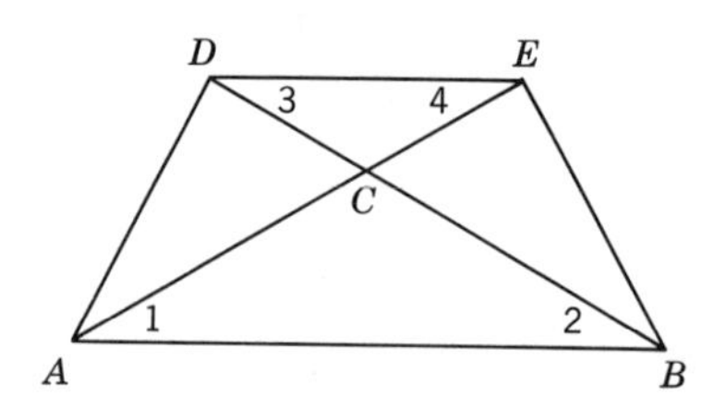

7. *Given:* Isosceles triangles:
$\triangle AFC$ with base AC.
$\triangle BGD$ with base BD.
$\triangle BHC$ with base BC.

Prove: $\triangle AED$ is isosceles.

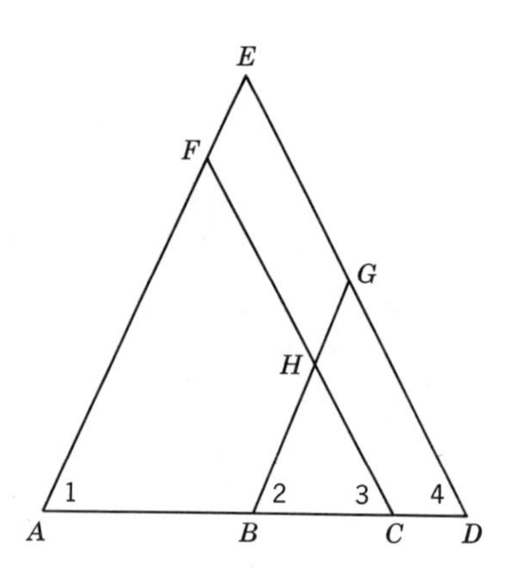

8. *Given:* Points A, B, C, and D lie in a line.
$\angle 1 = \angle D$.
$\angle 2 = \angle A$.
$CE = AE$.

Prove: $\triangle BCF$ is isosceles.

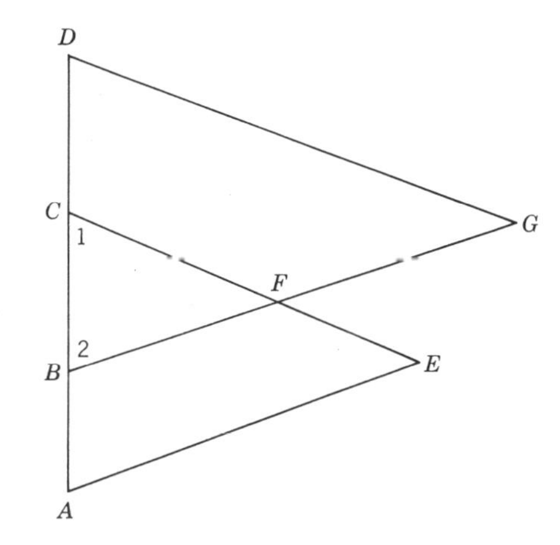

We are now in a position to prove a useful theorem concerning right triangles.

Definition 2.2 The side opposite the right angle in a right triangle is called the *hypotenuse* of the right triangle. The other two sides are sometimes called *arms* or *legs* of the right triangle.

Theorem 2.5

If the hypotenuse and an arm of one right triangle are equal, respectively, to the hypotenuse and an arm of a second right triangle, the right triangles are congruent.

Given: $\triangle DEF$ and $\triangle ABC$ are right triangles.
$DF = AC$.
$DE = AB$.

Prove: $\triangle DEF \cong \triangle ABC$.

Construction. On side AC of $\triangle ABC$ construct $\triangle D'E'F'$ so that $D'F' = DF$, $\angle 5 = \angle 6$, and $E'F' = EF$. Notice that by the construction $\triangle DEF \cong \triangle D'E'F'$. Draw line segment $E'B$.

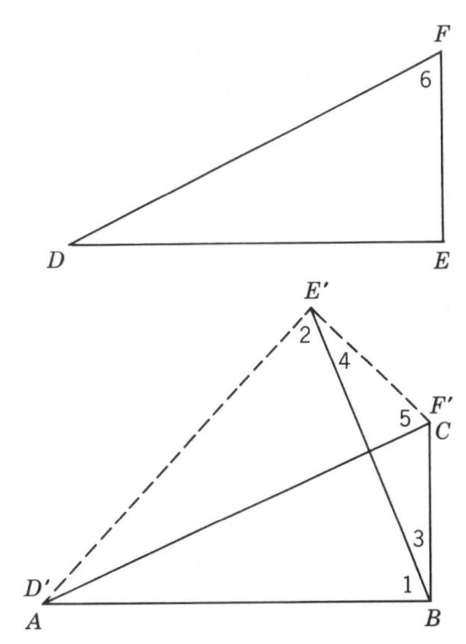

Proof:

STATEMENTS	REASONS
1. $\triangle DEF$ and $\triangle ABC$ are right triangles.	1. Given.
2. $DF = AC$; $DE = AB$.	2. Given.
3. $\triangle D'E'F'$ is constructed so that $D'F' = DF$, $\angle 5 = \angle 6$, and $E'F' = EF$.	3. Possible construction.
4. $\triangle D'E'F' \cong \triangle DEF$.	4. SAS = SAS.
5. $DE = D'E'$.	5. cpcte.
6. $AB = D'E'$.	6. Why?
7. $\angle 1 = \angle 2$.	7. Why?
8. $\angle ABC = \angle D'E'F'$.	8. Why?
9. $\angle 3 = \angle 4$.	9. Why?
10. $E'F' = BC$.	10. Why?
11. $\triangle ABC \cong \triangle D'E'F'$.	11. SAS = SAS.
12. $\triangle ABC \cong \triangle DEF$.	12. Why?

Theorem 2.5 gives us a fourth way of proving right triangles congruent. We could also prove right triangles congruent by the SSS, SAS, and ASA traditional methods. Our new method of proving congruence is sometimes abbreviated hs = hs.

EXERCISES

1. *Given:* $\triangle ABC$ is isosceles with base AB.
   $CD \perp AB$. perpindicular

   Prove: CD bis AB.

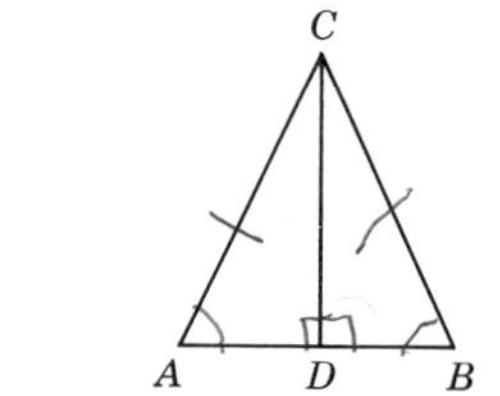

2. *Given:* ED and AC intersect at B.
   $AE \perp AC$.
   $CD \perp AC$.
   B midpoint of ED.
   $AE = CD$.

   Prove: B midpoint of AC.

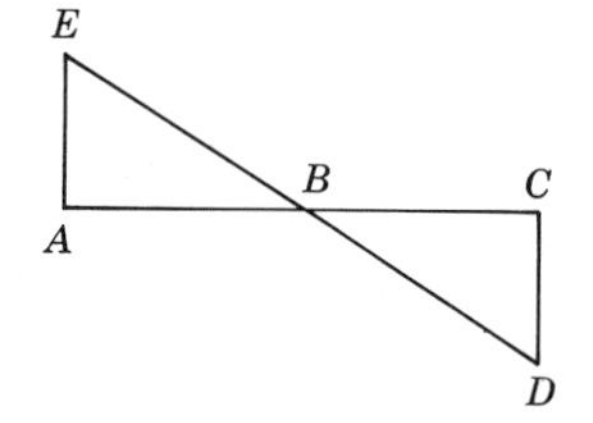

3. *Given:* $AC = BD$.
   $\angle CDA = \angle BAD = 90°$

   Prove: $AB = DC$.

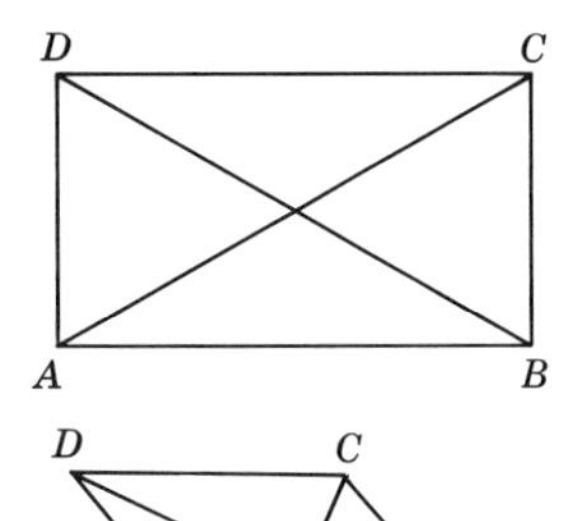

4. *Given:* DB and AC intersect at O.
   $\angle 1 = \angle 2 = 90°$.
   $CD = CB$.

   Prove: $OD = OB$.

5. Distinguish between **hypothesis** and **hypotenuse**.

6. Construct an isosceles right triangle with its two sides equal to AB.

7. *Given:* $\triangle ABC$ is isosceles with base AB.
   $BD \perp AC$
   $AE \perp BC$
   $\angle 1 = \angle 2$.

   Prove: $\angle 3 = \angle 4$.

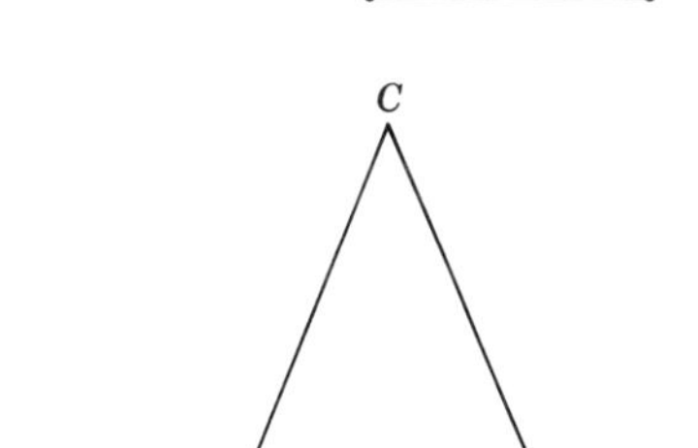

8. *Given:* $\triangle ABC$ and $\triangle ADE$ are isosceles with bases AB and AE, respectively. Line segment CD with $DF = CE$.
   $BF \perp CD$; $AD \perp DC$.

   Prove: $\angle 3 = \angle 4$.

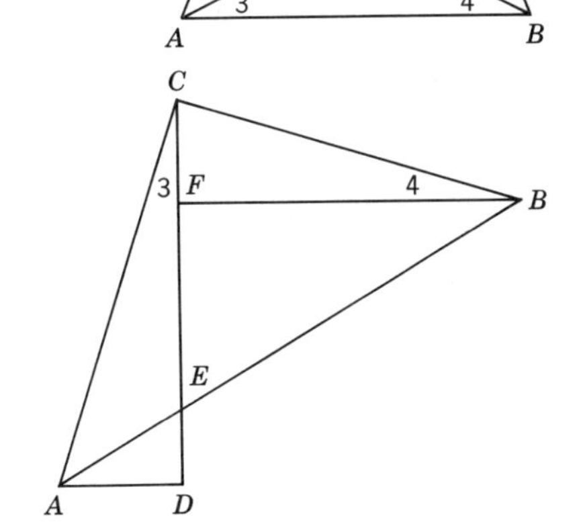

Review Test

Classify the following statements as true or false.

1. Two triangles are congruent if three sides of one triangle are equal, respectively, to three sides of a second triangle.

2. Two triangles are congruent if an angle and a side of one triangle are equal, respectively, to an angle and a side of a second triangle.
3. If two triangles are congruent, all corresponding parts of the two triangles are equal.
4. All equilateral triangles are congruent.
5. The vertex angle of an isosceles triangle can never equal one of the base angles.
6. The bisector of the vertex angle of an isosceles triangle bisects the base of the triangle.
7. A right triangle can never have more than one hypotenuse.
8. If two triangles have equal hypotenuses, the triangles are congruent.
9. If an angle appears to be a 45° angle in a drawing, it may be assumed the angle does have a measure of 45°.
10. If a triangle has two equal angles, it is isosceles.
11. If an angle in one triangle is equal to an angle in a second triangle, then the sides opposite these angles are equal.
12. The converse of a conditional statement is sometimes true.
13. An equilateral triangle is always equiangular.
14. A line segment has only one bisector.
15. The hypothesis is the clause following the word "if" in an "If–then" sentence.
16. If two triangles are congruent, then they are isosceles.
17. If two right triangles are congruent, each pair of corresponding acute angles are complementary.
18. Three pieces of information must always be known before the congruence of triangles can be shown.
19. The base of an isosceles triangle is always included between a pair of equal angles of the triangle.
20. If three angles of one triangle are equal, respectively, to three angles of a second triangle, the triangles are congruent.

3

Parallels and Parallelograms

In previous chapters, we have been concerned directly or indirectly with lines that intersected to form various angles. We will now center our attention on some lines that do not intersect.

Definition 3.1 Two lines are called *parallel lines* if, and only if, they are in the same plane and do not intersect.

If two lines do not intersect and do not lie in the same plane, they are called *skew* lines. However, in this book we will restrict our study to lines in the same plane.

We will need to know when two lines are parallel, and Definition 3.1 alone is not particularly useful. The next postulate will be more helpful.

Postulate 3.1 Two lines are parallel if they are both perpendicular to a third line.

When two lines are intersected by a third line in two distinct points, the third line is called a *transversal*. In Fig. 3.1, l_3 is a transversal intersecting lines

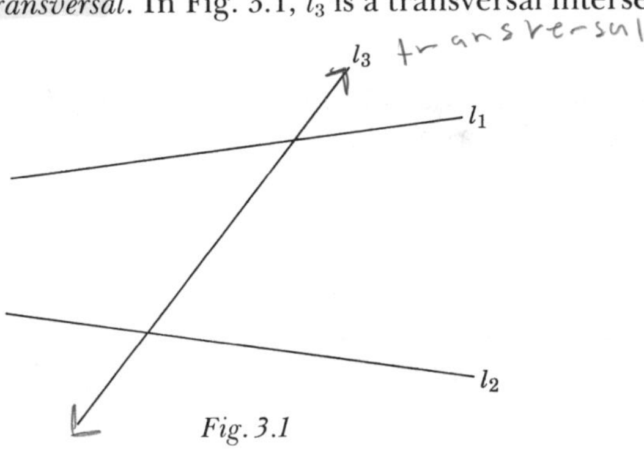

Fig. 3.1

l_1, and l_2. Several pairs of angles are formed by the transversal and the lines it intersects. We now make definitions to classify these pairs of angles.

Definition 3.2 If two lines are cut by a transversal, nonadjacent angles on opposite sides of the transversal but on the interior of the two lines are called *alternate interior angles*.

Definition 3.3 If two lines are cut by a transversal, angles on the same side of the transversal and in corresponding positions with respect to the lines are called *corresponding angles*.

Definition 3.4 If two lines are cut by a transversal, nonadjacent angles on opposite sides of the transversal and on the exterior of the two lines are called *alternate exterior angles*.

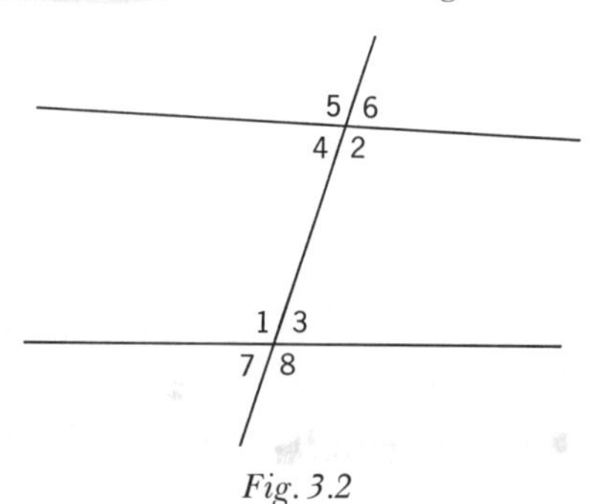

Fig. 3.2

In Fig. 3.2, $\angle 1$ and $\angle 2$ are a pair of alternate interior angles. Another pair of alternate interior angles is $\angle 3$ and $\angle 4$. $\angle 1$ and $\angle 5$, $\angle 3$ and $\angle 6$, $\angle 4$ and $\angle 7$, and $\angle 2$ and $\angle 8$ are examples of corresponding angles. $\angle 5$ and $\angle 8$ is one example of a pair of alternate exterior angles. Find another example.

A useful abbreviation for the word "parallel" is the symbol "$\|$". We are now ready to prove some useful theorems about parallel lines.

Theorem 3.1

If two lines are cut by a transversal so that alternate interior angles are equal, the lines are parallel.

Given: $\angle 3 = \angle 4$

Prove: $l_1 \| l_2$

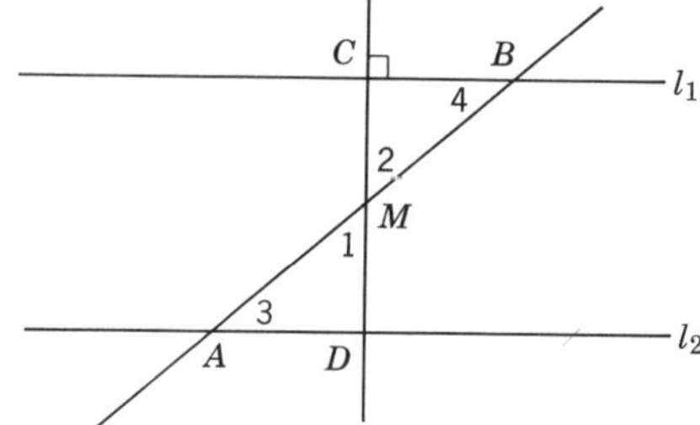

Construction: Construct the midpoint of AB and call it point M. Then construct a perpendicular to line l_1 passing through point M.

Proof:

STATEMENTS	REASONS
1. M is midpoint of AB.	1. Possible construction.
2. $MA = MB$.	2. Why?
3. $CD \perp l_1$.	3. Possible construction.
4. $\angle BCM = 90°$.	4. Why?
5. $\angle 1 = \angle 2$.	5. Why?
6. $\angle 3 = \angle 4$.	6. Given.
7. $\Delta ADM \cong \Delta BCM$.	7. ASA = ASA.
8. $\angle \mathrm{BCM} = \angle ADM$.	8. cpcte.
9. $\angle ADM = 90°$.	9. Why?
10. $CMD \perp l_2$.	10. Why?
11. $l_1 \parallel l_2$.	11. Why?

Theorem 3.1 gives us a procedure for constructing a line parallel to a given line through a point not on that given line. Suppose that we wish to construct a line parallel to a given line l that passes through a given point P which does not lie on line l. First we draw a line through the given point P that intersects the given line l. This line will be a transversal. We then copy $\angle 1$ in the position of $\angle 2$ using point p as vertex (Fig. 3.4, p. 44). Since the alternate interior angles are equal by our construction, Theorem 3.1 guarantees that the second side of $\angle 2$, line r, will be parallel to the given line l. This construction proves the existence of a parallel to a line through a point that is not on that line.

Postulate 3.2 Given a line l and a point P not on line l, there is only one line through point P that is parallel to line l.

Postulate 3.2, called the parallel postulate, is very important. Because it seems to be more complicated than most other postulates, many mathematicians in the past spent great amounts of time trying to prove it. It was a real breakthrough to discover that this postulate is independent from the other postulates, that is, could not be proved using the other postulates.

Certain other pairs of angles formed by a transversal may be used to show that lines are parallel, as the next two theorems indicate.

Fig. 3.3 Nikolai Ivanovich Lobachevsky (1793–1856). Lobachevsky, a Russian mathematician, was one of the first to recognize the merits of a geometry that denied Euclid's parallel postulate. In his geometry, more than one parallel can be drawn to a given line through a point not on that line. Unfortunately, his work in plane hyperbolic geometry was largely ignored by others until after his death. Non-euclidean geometries, such as Lobachevsky's, provided Albert Einstein with a basis for his theory of relativity. The work of Lobachevsky and others has made it apparent that Euclidean geometry is only one of many possible geometries. (Historical Picture Service, Chicago).

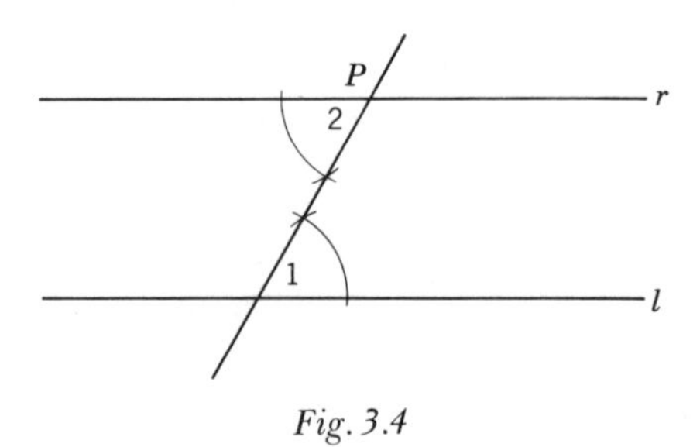

Fig. 3.4

Theorem 3.2

If two lines are cut by a transversal so that corresponding angles are equal, the lines are parallel.

Given: $\angle 1 = \angle 2$

Prove: $l_1 \parallel l_2$

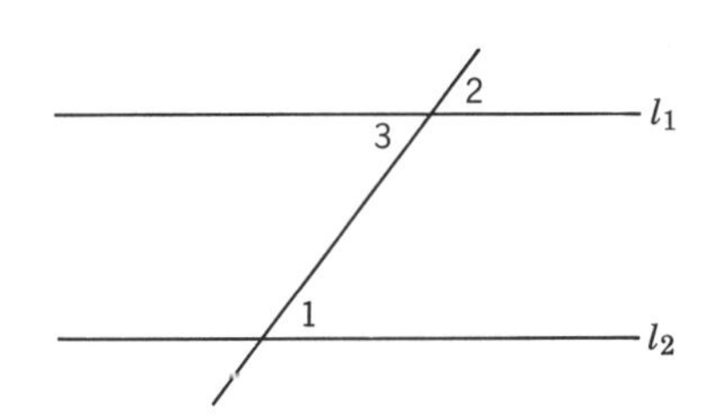

Proof:

STATEMENTS	REASONS
1. $\angle 1 = \angle 2$.	1. Why?
2. $\angle 3 = \angle 2$.	2. Why?
3. $\angle 1 = \angle 3$.	3. Why?
4. $l_1 \parallel l_2$.	4. Why?

Theorem 3.3

If two lines are cut by a transversal so that two interior angles on the same side of the transversal are supplementary, the lines are parallel.

The proof of this theorem is left as an exercise.

Theorem 3.4

If a line is perpendicular to one of two parallel lines, it is perpendicular to the other also.

Given: $l_1 \parallel l_2$
$l_3 \perp l_2$

Prove: $l_1 \perp l_3$

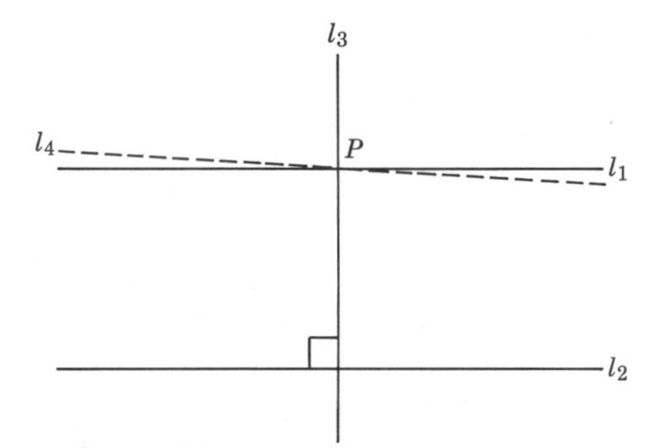

Plan. Construct $l_4 \perp l_3$ at point P and show that l_4 coincides with l_1.

Proof:

STATEMENTS	REASONS
1. Construct $l_4 \perp l_3$ at point P.	1. Possible construction.
2. $l_3 \perp l_2$.	2. Given.
3. $l_4 \parallel l_2$.	3. Two lines perpendicular to a third line are parallel.
4. $l_1 \parallel l_2$.	4. Given.
5. l_1 coincides with l_4.	5. Only one parallel to l_2 can exist through point P.
6. Since $l_3 \perp l_4$, then $l_3 \perp l_1$.	6. l_4 and l_1 are different names for the same line.

EXERCISES

1. Name two pairs of alternate interior angles.
2. Name two pairs of corresponding angles.
3. Prove Theorem 3.3.
4. *Prove:* if two lines are cut by a transversal so that alternate exterior angles are equal, the lines are parallel.
5. If $\angle 1 = \angle 2$, can ED intersect line AB?
6. If $\angle 1 = \angle 3$, can ED intersect line AB?
7. If $\angle 1 = \angle 3$, must ED intersect line AB?
8. If $\triangle ABC$ is isosceles and $\angle 1 = \angle 4$, can ED intersect line AB?
9. Draw a line on your paper about two inches long and mark a point about one inch from the line. Construct a parallel to the line passing through the given point. Can you devise a method of constructing the required parallel using Postulate 3.1?

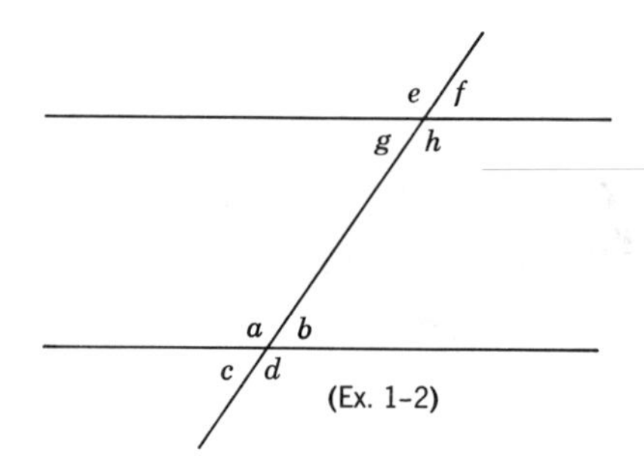

(Ex. 1-2)

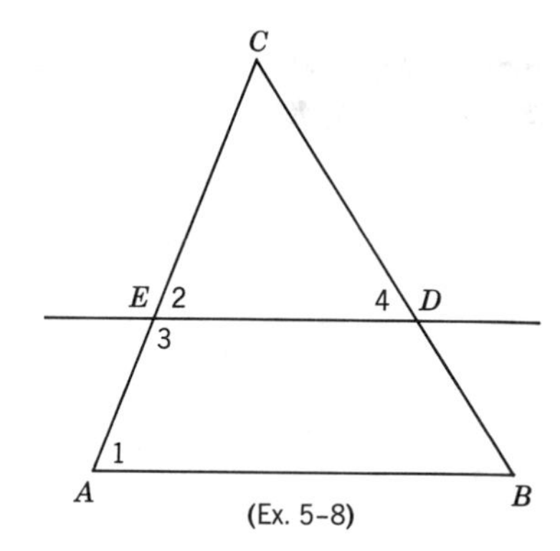

(Ex. 5-8)

10. *Given:* $\angle 2$ is supplementary to $\angle 3$.

Prove: $l_1 \parallel l_2$.

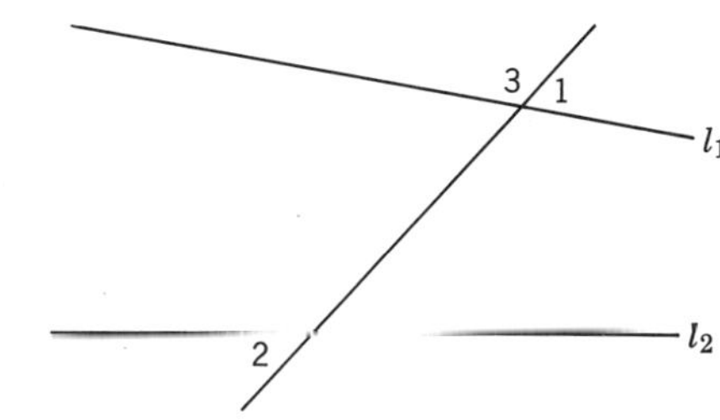

11. *Given:* $\angle 1$ and $\angle 2$ are supplementary.

Prove: $l_1 \parallel l_2$.

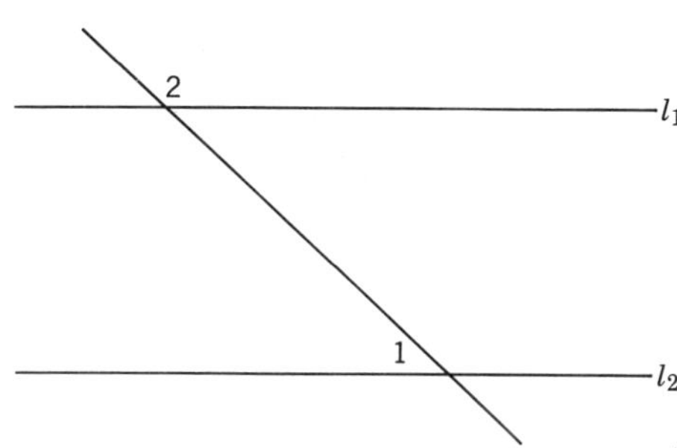

12. *Given:* AC and EB bisect each other at D.

Prove: $AE \parallel BC$.

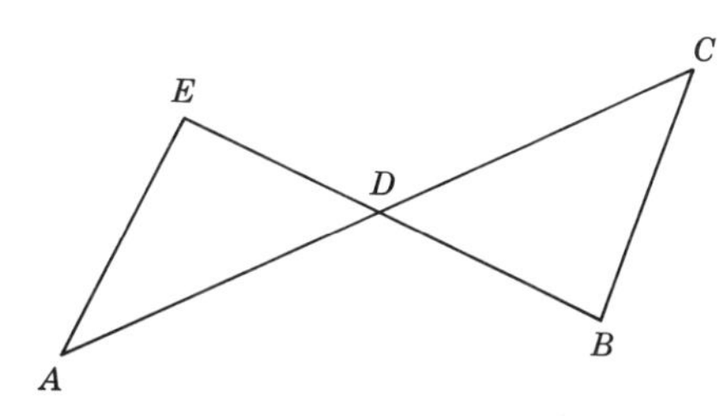

13. *Given:* $\triangle ABC$ is isosceles with base AB.
$\angle A = \angle 1$.

Prove: $AB \parallel ED$.

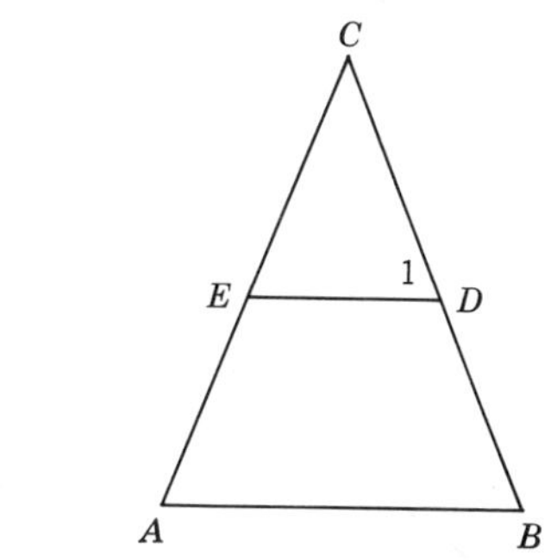

In the previous theorems and exercises we have always reasoned directly from whatever information was given to the required conclusion. There is another method of reasoning, called *indirect proof,* that we must consider before we continue. The method is illustrated by a simple example.

When John registered for the early morning geometry section, he knew he would have one of three possible instructors: Mr. Smith, Miss Jones, or Mr.

Schultz. Because he was curious, he made some inquiries to learn his teacher's name. He found out that Mr. Smith teaches only in the evenings and that Miss Jones quit to get married. John concluded that the only remaining possibility, Mr. Schultz, would be his teacher. By eliminating other possibilities, he reached this conclusion indirectly.

In *The Casebook of Sherlock Holmes*, the famous detective explains that his methods involve indirect proof:

> "That process", said I, "starts upon the supposition that when you have eliminated all which is impossible, then whatever remains, however improbable, must be the truth".

To eliminate possibilities we make use of the fact that geometry must be consistent. This means that a statement and its negation cannot both be true at the same time. If they were, we would have a contradiction that should be impossible if our work is sound. In any indirect proof, we will assume to be true the negation of what we are trying to prove. From this assumption, the given information and previous theorems, we try to find a contradiction. If we are lucky and this contradiction does develop, we know our assumption is false and can be eliminated. This leaves the statement we were trying to prove as the only remaining possibility.

If we wish to prove that two angles are equal by the indirect method, we could first assume that the two angles are not equal and show that this assumption leads to a contradiction. Similarly, if we wish to prove that two lines are parallel, we could assume that the lines are not parallel and show that this leads to a contradiction. If in each case we can eliminate the unwanted possibility, the desired conclusion is the only remaining choice.

We will use this method in the proof of Theorem 3.5, which is the converse of Theorem 3.1.

Theorem 3.5

If two parallel lines are cut by a transversal, all pairs of alternate interior angles are equal.

Given: $l_1 \parallel l_2$

Prove: $\angle 1 = \angle 2$

Plan. Use an indirect proof. Either $\angle 1 = \angle 2$ or $\angle 1 \neq \angle 2$. Assume $\angle 1 \neq \angle 2$ and show that this leads to a contradiction. Then $\angle 1 = \angle 2$ must be true.

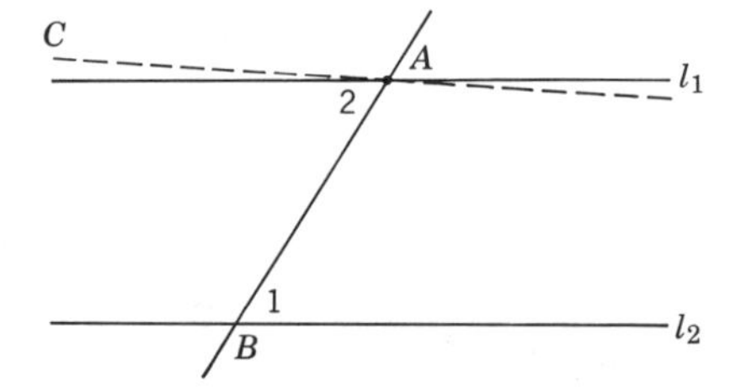

Proof:

STATEMENTS	REASONS
1. Assume $\angle 1 \neq \angle 2$.	1. Possible assumption.
2. At point A construct $\angle BAC = \angle 1$.	2. Possible construction.
3. $l_2 \parallel AC$.	3. If two lines are cut by a transversal so that alternate interior angles are equal, the lines are parallel.
4. $l_2 \parallel l_1$.	4. Given.
5. There are two lines parallel to l_2 through point A.	5. Steps 3 and 4.
6. Statement 5 is impossible.	6. There exists only one parallel to l_2 through point A.
7. $\angle 1 \neq \angle 2$ is false.	7. It leads to a contradiction.
8. $\angle 1 = \angle 2$.	8. It is the only remaining possibility.

The other pair of alternative interior angles can be proved equal by a similar argument.

Theorem 3.6

If two parallel lines are cut by a transversal, each pair of corresponding angles is equal.

Given: $l_1 \parallel l_2$

Prove: $\angle 1 = \angle 2$.

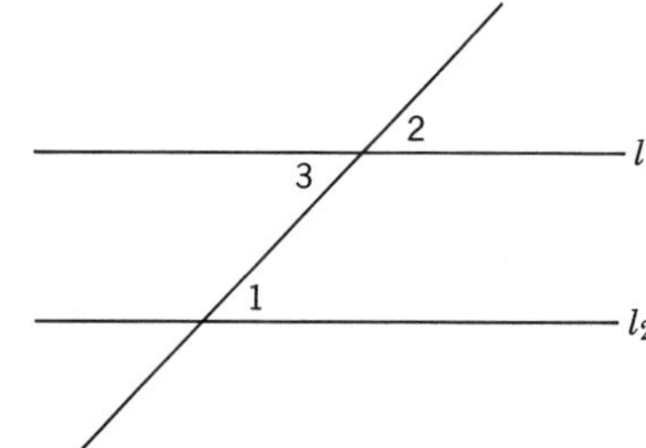

Proof:

STATEMENTS	REASONS
1. $l_1 \parallel l_2$.	1. Given.
2. $\angle 1 = \angle 3$.	2. Why?
3. $\angle 3 = \angle 2$.	3. Why?
4. $\angle 1 = \angle 2$.	4. Why?

Theorem 3.7

If two parallel lines are cut by a transversal, pairs of interior angles on the same side of the transversal are supplementary.

Given: $l_1 \parallel l_2$

Prove: $\angle 1$ supp. $\angle 2$

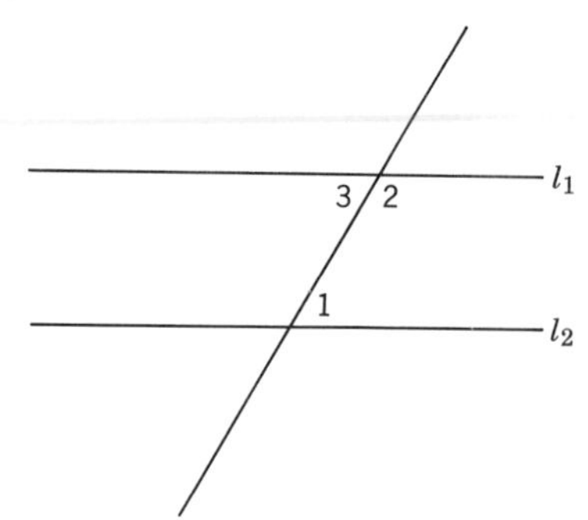

Proof:

STATEMENTS	REASONS
1. $l_1 \parallel l_2$.	1. Given.
2. $\angle 1 = \angle 3$.	2. Why?
3. $\angle 3$ supp. $\angle 2$.	3. Their sum is a straight angle.
4. $\angle 1$ supp. $\angle 2$.	4. A quantity may be substituted for its equal in any mathematical expression.

EXERCISES

1. Discuss the method of indirect proof in your own words.
2. If $l_1 \parallel l_2$, prove $\angle 1$ is supplementary to $\angle 2$.

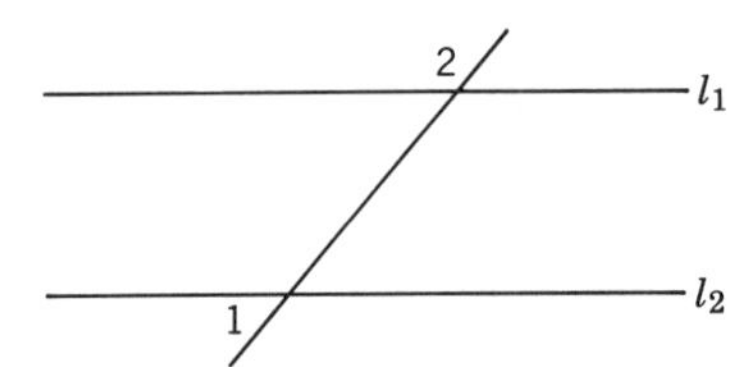

3. *Prove:* if two parallel lines are cut by a transversal, then all pairs of alternate exterior angles are equal.
4. If $l_1 \parallel l_2$ and $\angle 1 = 30°$, how many degrees are there in $\angle 2$? In $\angle 3$?

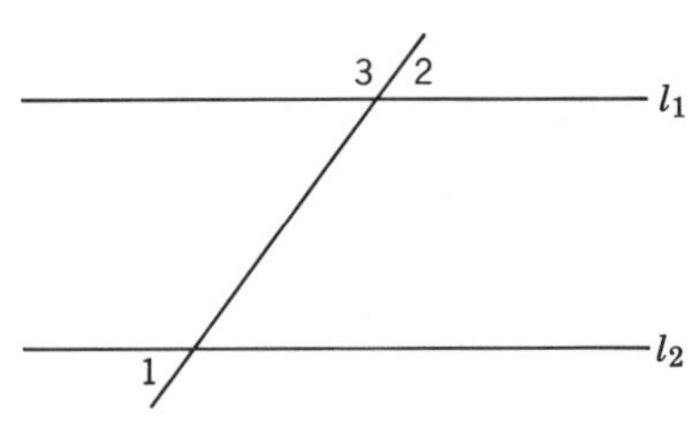

5. *Given:* $AB \parallel DC$ and $AB = CD$.

   Prove: $\angle A = \angle C$.

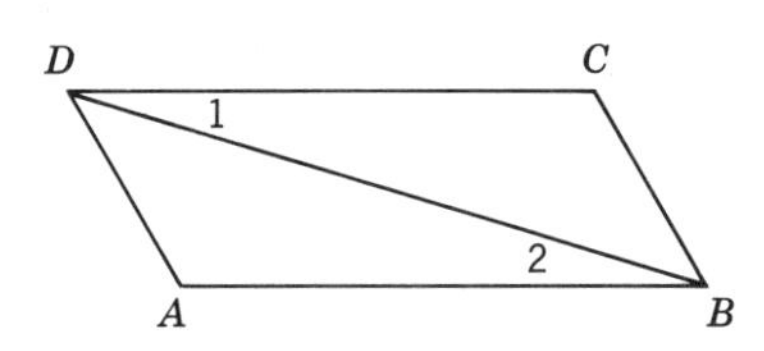

6. If line AB is parallel to line CD and line EF is parallel to line GH, prove that $\angle 1 = \angle 2$.

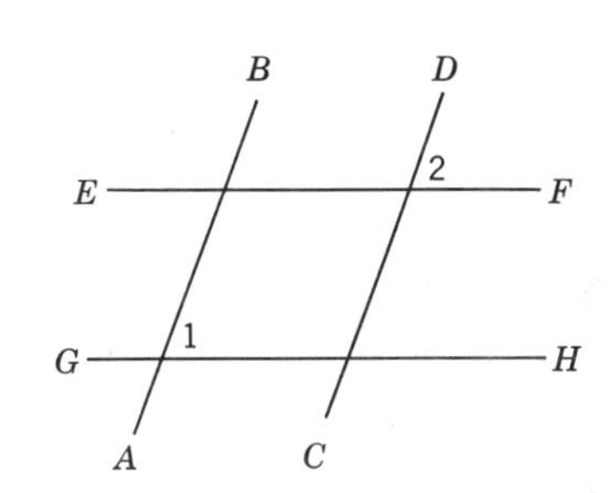

7. If line AB is parallel to line CD and $BC = DC$, prove that line BD bisects $\angle CBA$.

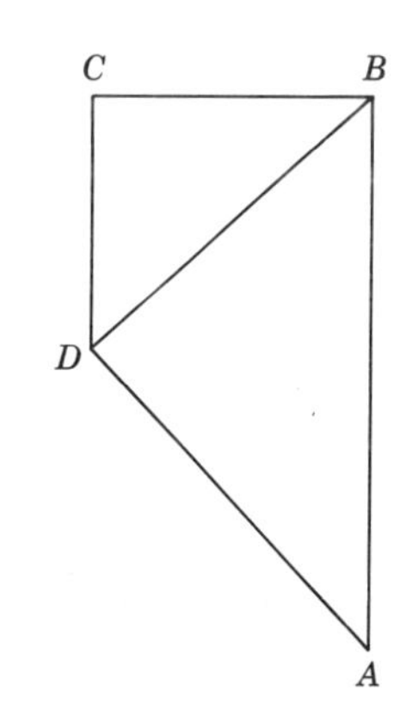

8. *Given:* AD and BC intersect at E.
   $AB \parallel CD$.
   $CE = DE$.

   Prove: $AE = BE$.

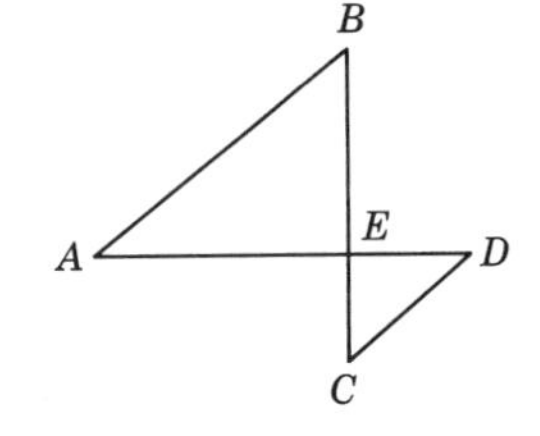

9. *Given:* $\triangle ABE$.
   $AB \parallel CD$.
   $CE = DE$.

   Prove: $AE = BE$.

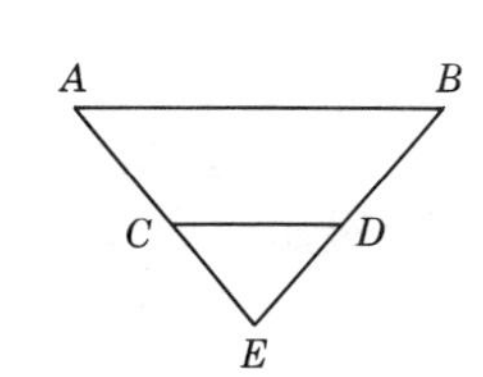

10. *Given:* $AD \parallel BE$, $BD \parallel CE$ and B is midpoint of AC.

    Prove: $BE = AD$.

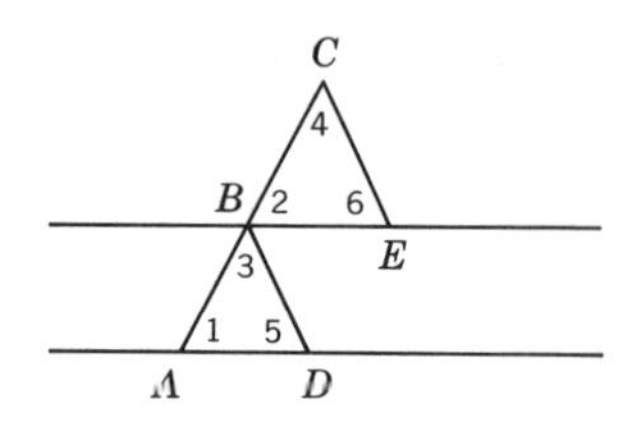

We can now prove one of the most important theorems we encounter in this course. We will also prove many corollaries to this theorem. A *corollary* is an easy theorem to prove because it is an immediate consequence of a major theorem.

Theorem 3.8

The sum of the measures of the interior angles of a triangle is 180°.

Given: $\triangle ABC$

Prove: $\angle A + \angle B + \angle 2 = 180°$

Construction: Construct line l_1 through C parallel to AB.

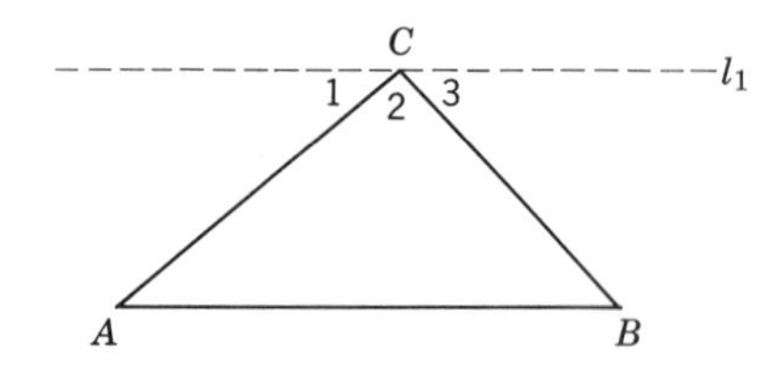

Proof:

STATEMENTS	REASONS
1. Line l_1 through C is parallel to AB.	1. By construction.
2. $\angle 1 + \angle 2 + \angle 3$ is a straight angle.	2. The whole is equal to the sum of its parts.
3. $\angle 1 + \angle 2 + \angle 3 = 180°$.	3. A straight angle equals 180°.
4. $\angle 1 = \angle A$.	4. If two parallel lines are cut by a transversal, the alternate interior angles are equal.
5. $\angle 3 = \angle B$.	5. Why?
6. $\angle A + \angle B + \angle 2 = 180°$.	6. Why?

Corollary 3.9 A triangle can have at most one right or obtuse angle.

The proof is indirect. Assume a triangle has more than one right angle. Then the sum of its angles must be greater than 180°, which contradicts Theorem 3.8. Therefore, our assumption is false and a triangle has at most one right angle. The case for the obtuse angle is left as an exercise.

Corollary 3.10 The acute angles of a right triangle are complementary.

The proof of this corollary is left as an exercise.

Corollary 3.11 If two angles of one triangle are equal, respectively, to two angles of a second triangle, their third angles are equal.

Given: $\angle A = \angle D$
$\angle C = \angle F$

Prove: $\angle B = \angle E$.

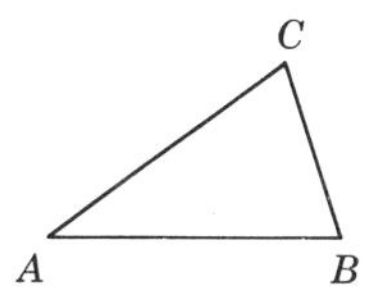

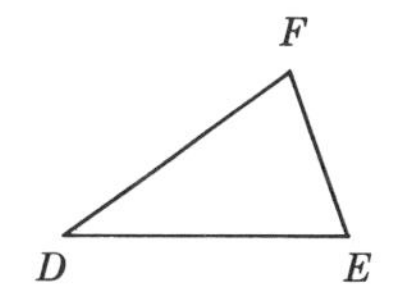

STATEMENTS	REASONS
1. $\angle A = \angle D$. $\angle C = \angle F$	1. Given.
2. $\angle A + \angle B + \angle C = 180°$. $\angle D + \angle E + \angle F = 180°$.	2. Why?
3. $\angle A + \angle B + \angle C = \angle D + \angle E + \angle F$.	3. Why
4. $\angle B = \angle E$.	4. Equals subtracted from equals are equal.

Corollary 3.12 If two angles and any side of one triangle are equal, respectively, to two angles and any side of a second triangle, the triangles are congruent. (AAS = AAS).

The proof of this corollary is left as an exercise. Notice that Corollary 3.12 gives us another way to prove that two triangles are congruent. We now have five ways to prove triangles congruent: SSS, SAS, ASA, hs, and AAS.

Definition 3.5 If one of the sides of a triangle is extended, the angle formed which is adjacent to an angle of the triangle is called an *exterior angle*.

In the figure, angles a, b, and c are exterior angles.

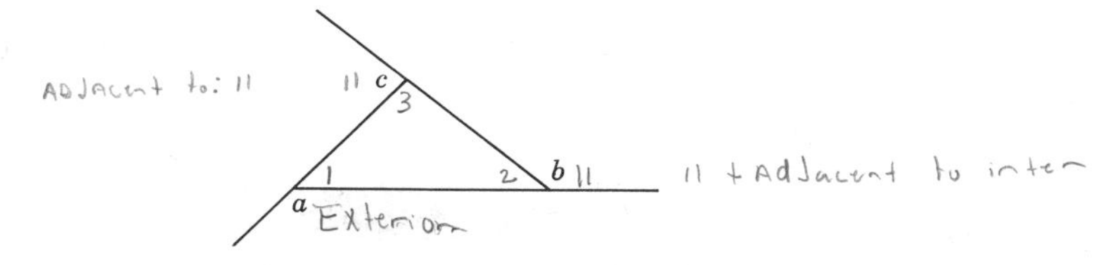

The following is a very important theorem about exterior angles of a triangle.

Corollary 3.13 An exterior angle of a triangle is equal to the sum of the nonadjacent interior angles.

Given: $\triangle ABC$ with $\angle A$, $\angle C$, and $\angle 3$.

Prove: $\angle 1 = \angle A + \angle C$

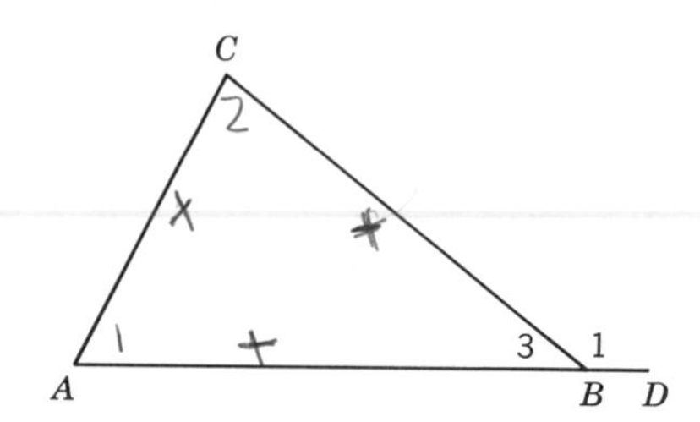

Proof:

STATEMENTS	REASONS
1. $\triangle ABC$ with $\angle A$, $\angle C$ and $\angle 3$.	1. Given.
2. $\angle A + \angle C + \angle 3 = 180°$.	2. The sum of the angles of a triangle equals 180°.
3. $\angle 1$ is supplementary to $\angle 3$.	3. Definition of supplementary angles.
4. $\angle 1 + \angle 3 = 180°$.	4. Why?
5. $\angle A + \angle C + \angle 3 = \angle 1 + \angle 3$.	5. Why?
6. $\angle A + \angle C = \angle 1$.	6. Equals subtracted from equals are equal.

Theorem 3.14

If the hypotenuse and an acute angle of one right triangle are equal, respectively, to the hypotenuse and an acute angle of a second right triangle, the triangles are congruent.

The proof of this theorem is left as an exercise. Notice that we now have a sixth method of proving triangles congruent. Theorem 3.14 is often abbreviated as ha = ha.

EXERCISES

1. *Prove:* a triangle can have at most one obtuse angle.
2. Prove Corollary 3.10.
3. Prove Corollary 3.12.
4. Prove Theorem 3.14.

5. Given that $AB \| DC$ and $AD \| BC$, what are the measures of $\angle B$, $\angle 1$, $\angle D$, and $\angle 2$?

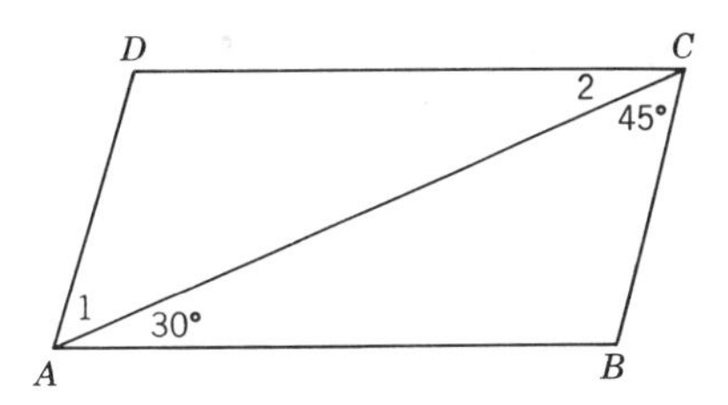

6. *Given:* $\triangle ABC$
   $AC = BC$
   $\angle 1 = \angle 2$

   Prove: $\angle 3 = \angle 4$.

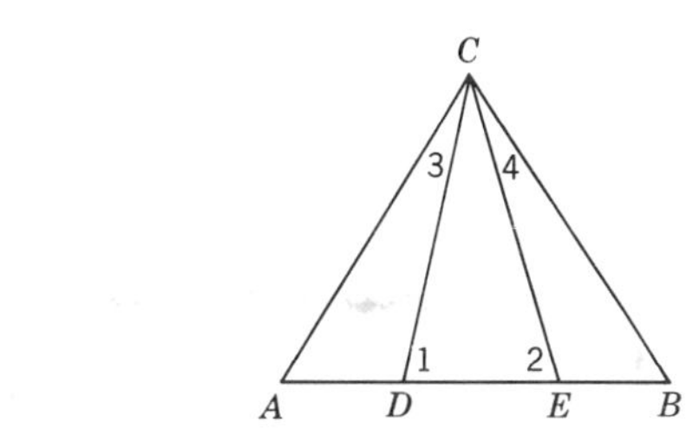

7. *Given:* AC and ED intersect at B
   $AE \perp AC$, $CD \perp AC$
   and $BE = BD$.

   Prove: $AB = CB$

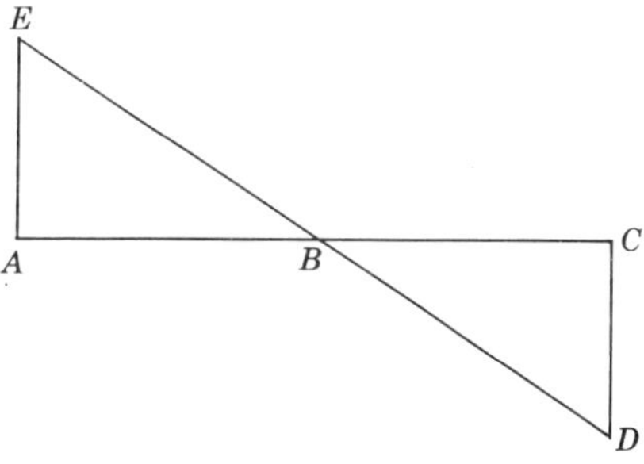

8. *Given:* AC is perpendicular to line BD, $\angle 1 = \angle 2$.

   Prove: AC bisects BD.

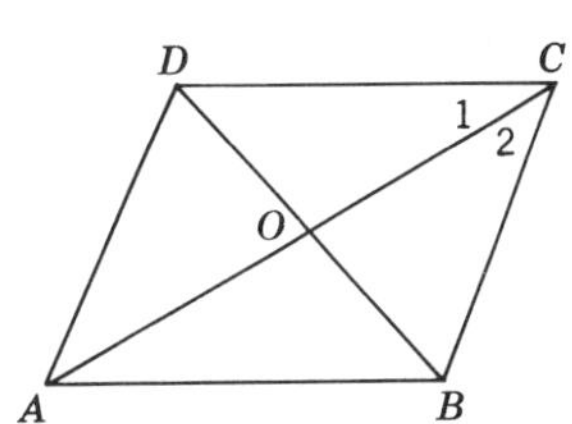

9. *Prove:* if two exterior angles of a triangle are equal, then the triangle is isosceles.
10. What are the measures of the angles of an equilateral triangle? Of an isosceles right triangle?
11. *Prove:* if the sum of two angles of a triangle equals the third angle, then the triangle is a right triangle.

A figure is called a *simple figure* if no pair of nonadjacent sides intersects. A *diagonal* of a figure is a line segment that joins two nonadjacent vertices of the figure. If any diagonal of a simple figure is partly in the exterior of the figure, the figure is called *concave*. Otherwise the figure is called *convex*. In this course we will restrict our discussion to convex figures.

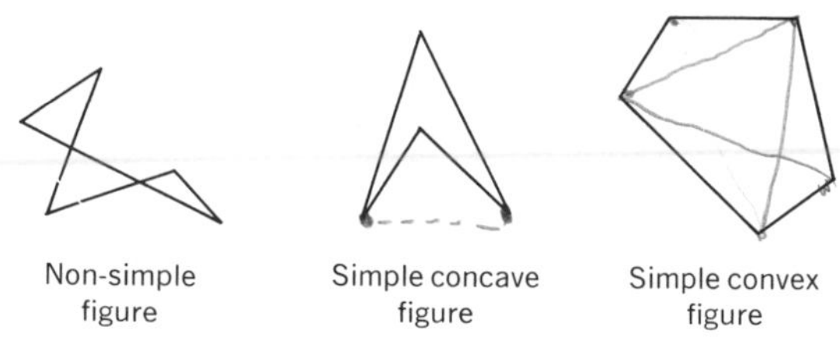

Fig. 3.5

Definition 3.6 A *polygon* is a convex figure with many angles.

Postulate 3.3 A polygon has the same number of sides as angles.

There are some useful theorems about the angles of a polygon that follow easily from Theorem 3.8.

Theorem 3.15

The sum of the interior angles of a polygon is given by the formula $S = (n-2) \cdot 180°$, where n is the number of sides of the polygon.

Given: A polygon with n sides.

Prove: The sum of the interior angles of the polygon equals $(n-2) \cdot 180°$. $[S = (n-2) \cdot 180°]$

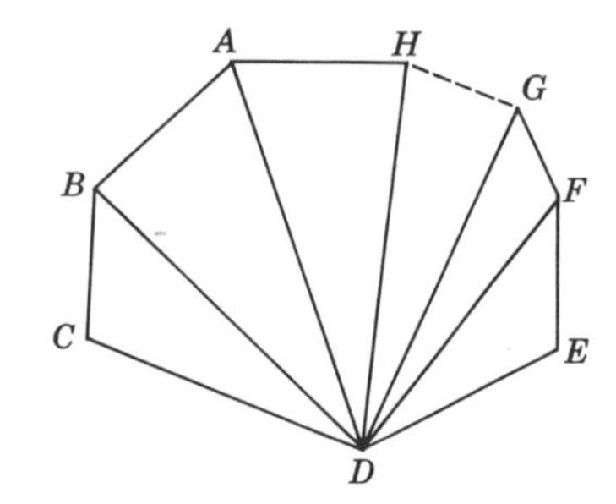

Proof: Pick some vertex. We will choose vertex D. Draw all possible diagonals in the polygon from vertex D. We note that it is impossible to draw diagonals to vertex C or vertex E because these vertices are adjacent to vertex D. We can draw $(n-3)$ diagonals; one to each vertex except vertices C, D, and E. Diagonal DB will form triangle BDC; diagonal DA will form triangle ADB; diagonal DH will form triangle HDA; and so on until the last diagonal DF forms two triangles: triangle FDG and triangle EDF. Therefore, $(n-2)$ triangles will be formed. The sum of the angles of each triangle is 180°, and the sum of the angles of $(n-2)$ triangles is $(n-2) \cdot 180°$. But the sum of the angles of all $(n-2)$ triangles is equal to the sum of the angles of the polygon. Hence, the sum of the angles of the polygon is equal to $(n-2) \cdot 180°$.

Notice that we did not mention a particular number of sides; the proof will apply to all polygons regardless of the number of sides.

Theorem 3.16

The sum of the exterior angles of a polygon is always 360°.

Given: A polygon with n sides.

Prove: The sum of the exterior angles equal 360°.

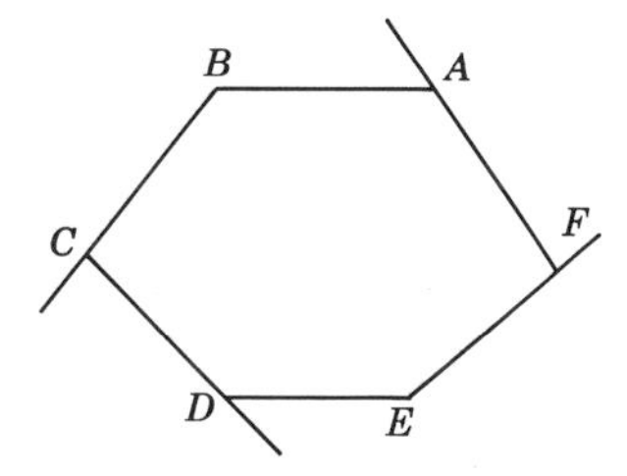

Proof: There are n straight angles through the n vertices of the polygon. Each of these n straight angles is the sum of an interior angle and an adjacent exterior angle. We know that the measure of n straight angles equals $n \cdot 180°$ and that the sum of the interior angles is $(n-2) \cdot 180°$. If we called the sum of the exterior angles E, we have

$$E + (n-2) \cdot 180° = 180°n$$
$$E + 180°n - 360° = 180°n$$
$$E - 360° = 0$$
$$E = 360°$$

Definition 3.7 An *equilateral polygon* is a polygon all of whose sides are of equal measure.

Definition 3.8 An *equiangular polygon* is a polygon all of whose angles are of equal measure.

Definition 3.9 A *regular polygon* is a polygon that is both equilateral and equiangular.

Theorem 3.17

The measure of an interior angle, a, of a regular polygon is given by the formula:

$$a = \frac{(n-2) \cdot 180°}{n}$$

where n is the number of sides of the polygon.

Proof: We know the sum of the measures of the interior angles of any polygon is given by the formula $S = (n-2) \cdot 180°$ and that there are n equal angles in an n-sided regular polygon. Therefore, each angle is equal to $1/n$ times the sum of the angles. Hence,

$$a = \frac{S}{n} = \frac{(n-2) \cdot 180°}{n}$$

EXERCISES

1. Find the sum of the interior angles of an octagon (an eight-sided polygon).
2. Find the sum of the interior angles of a decagon (a ten-sided polygon).
3. The sum of the interior angles of a polygon is 2880°. Find the number of sides of the polygon.
4. Find the sum of the exterior angles of a hexagon (a six-sided polygon).
5. Find the measure of one interior angle of a regular pentagon (a five-sided polygon).
6. The measure of an interior angle of a regular polygon is 150°. How many sides does the polygon have?
7. Can the measure of an interior angle of a regular polygon equal 115°? Why?
8. Can the sum of the interior angles of a polygon equal 2000°? Why?
9. Can an exterior angle of any regular polygon equal 15°?
10. Find the measure of an exterior angle of a regular polygon with 15 sides.
11. Show that the formula, $a = 180° - 360°/n$, is equivalent to the formula given in theorem 3.17.
12. In a certain regular polygon, twice an exterior angle equals one interior angle. How many sides does the polygon have?
13. The number of degrees in an exterior angle of a certain regular polygon is ten times the number of sides. How many sides does the polygon have?
14. If the number of sides of a certain polygon were doubled, the sum of the interior angles would increase by 900°. How many sides does the first polygon have?
15. *Prove:* if the sum of the exterior angles of a polygon equals the sum of the interior angles, the polygon is a four-sided polygon.

Theorem 3.18

If two lines are cut by a transversal so that alternate interior angles are not equal, the lines are not parallel.[1]

Given: $\angle 1 \neq \angle 2$.

Prove: $l_1 \nparallel l_2$

Plan: Use an indirect proof and assume $l_1 \parallel l_2$ and show this leads to a contradiction.

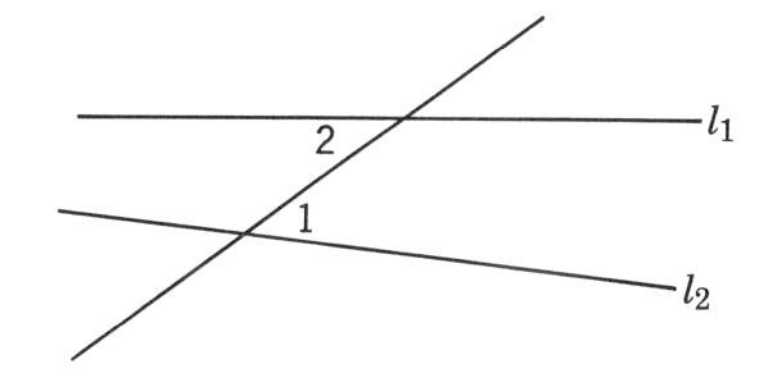

Proof:

STATEMENTS	REASONS
1. $l_1 \parallel l_2$.	1. Assumption.
2. $\angle 1 = \angle 2$.	2. If two parallel lines are cut by a transversal, pairs of alternate interior angles are equal.
3. $\angle 1 \neq \angle 2$.	3. Given.
4. $l_1 \parallel l_2$ is false.	4. It leads to a contradiction in statements 2 and 3.
5. $l_1 \nparallel l_2$.	5. It is the only remaining possibility.

Theorem 3.19

If two lines are cut by a transversal so that corresponding angles are not equal, the lines are not parallel.

Given: $\angle 1 \neq \angle 2$

Prove: $l_1 \nparallel l_2$

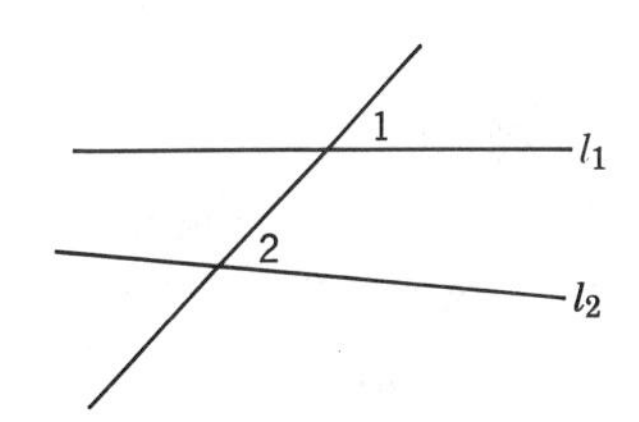

[1]Theorem 3.18 has a special relationship to Theorem 3.5 and is called the contrapositive of Theorem 3.5. Because of this relationship, the two theorems are equivalent statements, and Theorem 3.18 is not a new result. We will study contrapositives of statements in more detail in Chapter 10. The proof of Theorem 3.18 is included only as an example of indirect reasoning.

Proof 1 (indirect):

STATEMENTS	REASONS
1. $l_1 \parallel l_2$.	1. Assumption.
2. $\angle 1 = \angle 2$.	2. Why?
3. $\angle 1 \neq \angle 2$.	3. Why?
4. $l_1 \parallel l_2$ is false.	4. Why?
5. $l_1 \nparallel l_2$.	5. Why?

Given: $\angle 1 \neq \angle 2$

Prove: $l_1 \nparallel l_2$

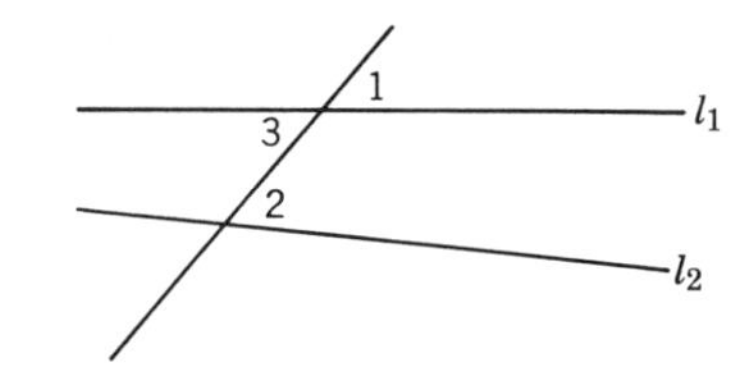

Proof 2 (direct):

STATEMENTS	REASONS
1. $\angle 1 = \angle 3$.	1. Vertical angles are equal.
2. $\angle 1 \neq \angle 2$.	2. Given.
3. $\angle 3 \neq \angle 2$.	3. A quantity may be substituted for its equal in any mathematical expression.
4. $l_1 \nparallel l_2$.	4. Theorem 3.18.

Theorem 3.20

If two lines are cut by a transversal so that two interior angles on the same side of the transversal are not supplementary, the lines are not parallel.

Theorem 3.21

If two nonparallel lines are cut by a transversal, the pairs of alternate interior angles are not equal.

The proofs of Theorems 3.20 and 3.21 are left as exercises.

EXERCISES

1. Prove Theorem 3.20.
2. Prove Theorem 3.21.

3. *Prove:* if two nonparallel lines are cut by a transversal, pairs of corresponding angles are not equal.
4. *Prove:* if two nonparallel lines are cut by a transversal, angles on the same side of the transversal and on the interior of the lines are not supplementary.
5. *Prove:* if the sum of the angles of a polygon is not 180°, the polygon is not a triangle.

*6. *Prove:* two distinct lines parallel to a third line are parallel to each other.

*7. *Prove:* two lines intersect in at most one point.

Until now, we have dealt principally with triangles and their properties. We now broaden our outlook by considering other important geometric figures.

Definition 3.10 A *quadrilateral* is a polygon with four sides.

Definition 3.11 A *parallelogram* is a quadrilateral whose opposite sides are parallel. A convenient symbol for parallelogram is "▱".

Theorem 3.22

A diagonal of a parallelogram divides the parallelogram into two congruent triangles.

Given: $ABCD$ is a parallelogram.

Prove: $\triangle ABD \cong \triangle CDB$.

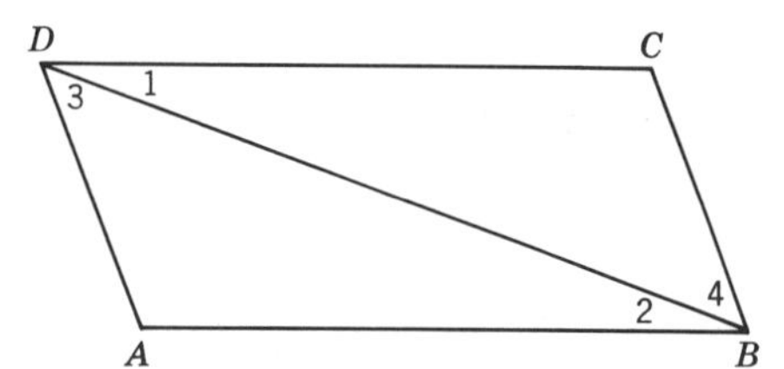

Proof:

STATEMENTS	REASONS
1. $ABCD$ is a parallelogram.	1. Given.
2. $AB \parallel DC$; $AD \parallel BC$.	2. The opposite sides of a parallelogram are parallel.
3. $\angle 1 = \angle 2$; $\angle 3 = \angle 4$.	3. If two parallel lines are cut by a transversal, then alternate interior angles are equal.
4. $BD = DB$.	4. Why?
5. $\triangle ABD \cong \triangle CDB$.	5. Why?

Corollary 3.23 Opposite sides of a parallelogram are equal.

Corollary 3.24 Nonconsecutive angles of a parallelogram are equal.

Theorem 3.25

Consecutive angles of a parallelogram are supplementary.

The proofs of Corollaries 3.23 and 3.24 and Theorem 3.25 are left as exercises.

Definition 3.12 The *distance from a point to a line* is the measure of the perpendicular segment drawn from the point to the line.

Theorem 3.26

Parallel lines are always the same distance apart.

Given: $l_1 \parallel l_2$
$AB \perp l_2$ and AB is the distance from point A to line l_2.
$CD \perp l_2$ and CD is the distance from point C to line l_2.

Prove: l_1 and l_2 are always the same distance apart.

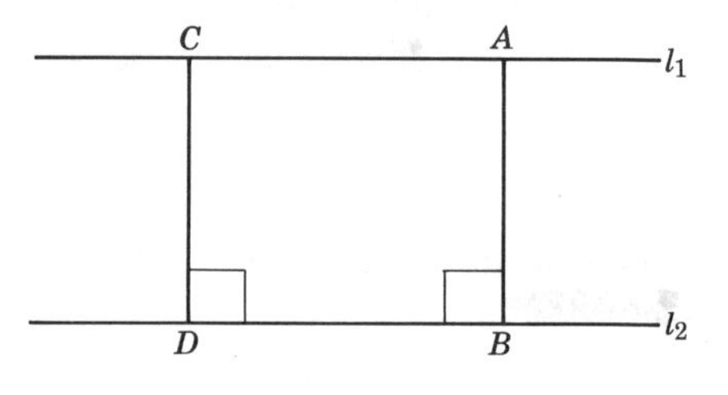

Plan. Show that two arbitrary points on line l_1 are equidistant from line l_2. Since the points are arbitrary, all points on line l_1 are equidistant from line l_2. Hence, l_1 and l_2 are always the same distance apart.

Proof:

STATEMENTS	REASONS
1. $l_1 \parallel l_2$.	1. Given.
2. $AB \perp l_2$; $CD \perp l_2$.	2. Given.
3. $CD \parallel AB$.	3. Two lines perpendicular to a third line are parallel.
4. $CDBA$ is a parallelogram.	4. Why?
5. $CD = AB$.	5. Why?
6. l_1 and l_2 are always the same distance apart.	6. Why?

Theorem 3.27

If both pairs of opposite sides of a quadrilateral are equal, then the quadrilateral is a parallelogram.

Given: $AB = CD$.
$AD = CB$.

Prove: $ABCD$ is a parallelogram.

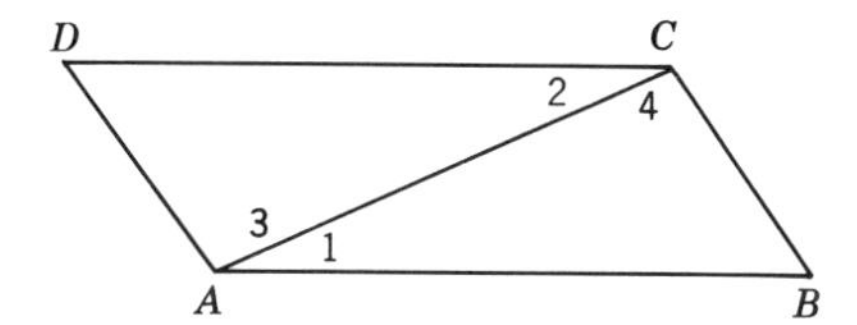

Proof:

STATEMENTS	REASONS
1. $AB = CD$. $AD = CB$.	1. Given.
2. $CA = AC$.	2. Why?
3. $\triangle ABC \cong \triangle CDA$.	3. Why?
4. $\angle 1 = \angle 2$; $\angle 3 = \angle 4$	4. cpcte.
5. $AB \parallel DC$; $AD \parallel BC$.	5. Why?
6. $ABCD$ is a parallelogram.	6. Why?

Theorem 3.28

If two opposite sides of a quadrilateral are both parallel and equal, the quadrilateral is a parallelogram.

The proof of Theorem 3.28 is left as an exercise.

Theorem 3.29

The diagonals of a parallelogram bisect each other.

Given: $ABCD$ is a parallelogram.

Prove: DB and AC bisect each other.

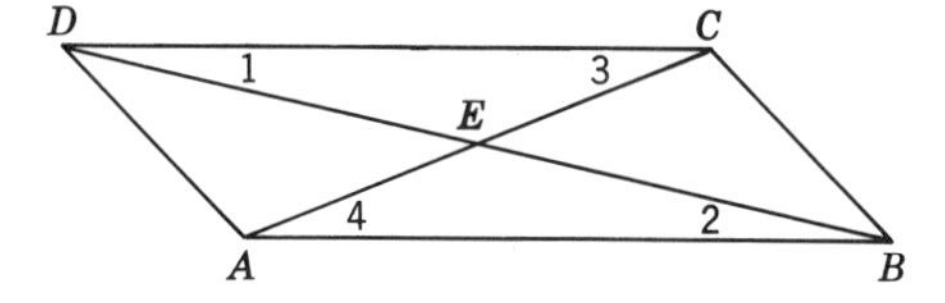

Proof:

STATEMENTS	REASONS
1. $ABCD$ is a parallelogram.	1. Given.
2. $DC \parallel AB$.	2. Why?

STATEMENTS	REASONS
3. $\angle 1 = \angle 2$; $\angle 3 = \angle 4$.	3. Why?
4. $DC = BA$.	4. Opposite sides of a parallelogram are equal.
5. $\triangle DCE \cong \triangle BAE$.	5. ASA = ASA.
6. $AE = CE$; $BE = DE$.	6. cpcte.
7. AC and BD bisect each other.	7. Why?

Theorem 3.30

If the diagonals of a quadrilateral bisect each other, then the quadrilateral is a parallelogram.

Given: DB and AC bisect each other at point E.

Prove: $ABCD$ is a parallelogram.

Proof:

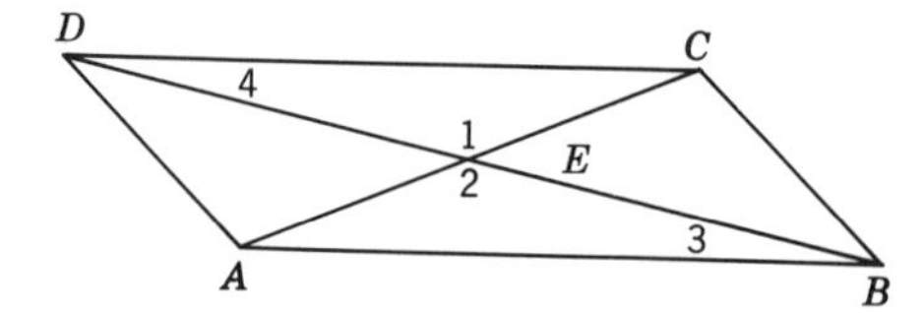

STATEMENTS	REASONS
1. DB and AC bisect each other at point E.	1. Given.
2. $DE = BE$; $CE = AE$.	2. A bisected line segment is divided into two equal parts.
3. $\angle 1 = \angle 2$.	3. Why?
4. $\triangle DCE \cong \triangle BAE$.	4. SAS = SAS.
5. $\angle 3 = \angle 4$.	5. cpcte.
6. $DC \parallel BA$.	6. Why?
7. $DC = BA$.	7. cpcte.
8. $ABCD$ is a parallelogram.	8. If two opposite sides of a quadrilateral are both parallel and equal, the quadrilateral is a parallelogram.

EXERCISES

1. Prove Corollary 3.23.
2. Prove Corollary 3.24.
3. Prove Theorem 3.25.

4. Prove Theorem 3.28.

5. *Prove:* if all pairs of consecutive angles of a quadrilateral are supplementary, then the quadrilateral is a parallelogram.

6. $ABCD$ is a parallelogram with E and F midpoints of AB and DC, respectively. Prove that $EBFD$ is a parallelogram.

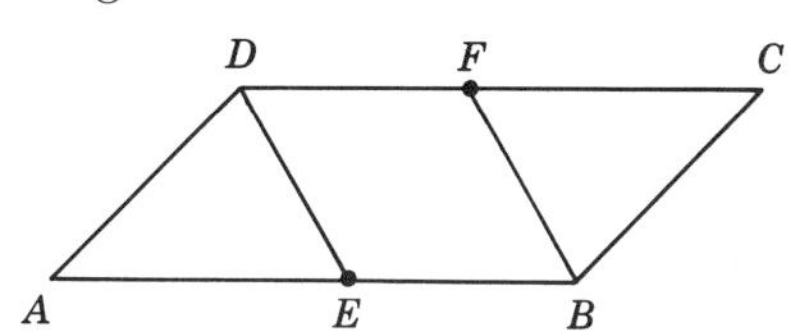

7. If one angle of a certain parallelogram is twice another, find the measure of each of its angles.

8. *Given:* $\square ABCD$, LN bisects diagonal AC at M.

   Prove: AC bisects LN.

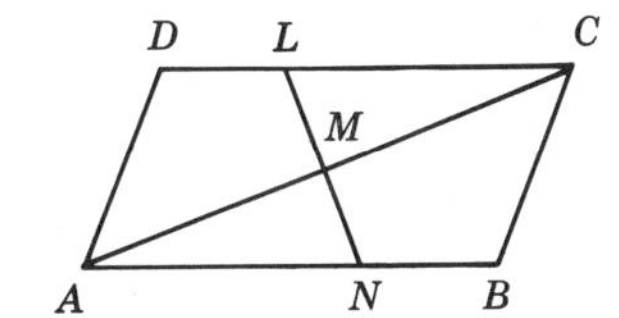

9. *Given:* Quadrilateral $ABCD$, with $AB \parallel DC$, and $\angle A = \angle C$.

   Prove: $ABCD$ is a parallelogram.

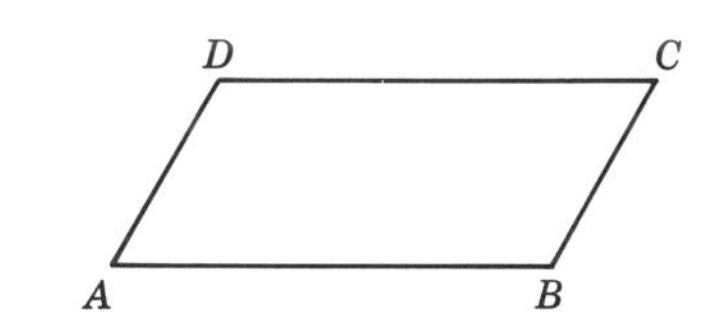

10. *Given:* Quadrilateral $ABCD$, with diagonals intersecting at M.
    $\angle 1 = \angle 2$.
    $AB = DC$.

    Prove: $MD = MB$.

11. *Given:* $\square ABCD$, with diagonal AC.
    $AM = NC$.

    Prove: $MBND$ is a parallelogram.

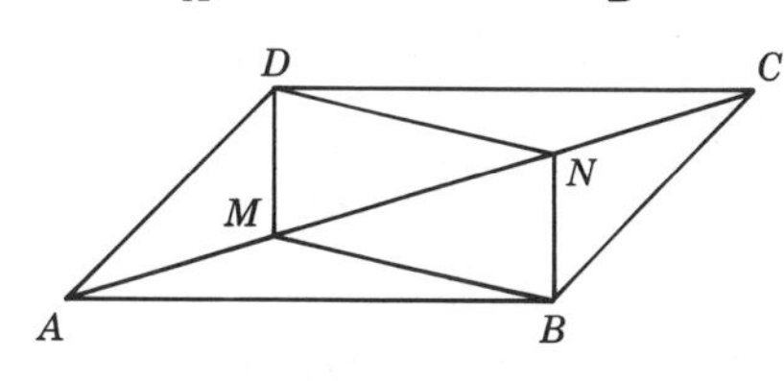

12. *Given:* $\square ABCD$.
    $AM \perp DB$.
    $CN \perp DB$.

    Prove: $AMCN$ is a parallelogram.

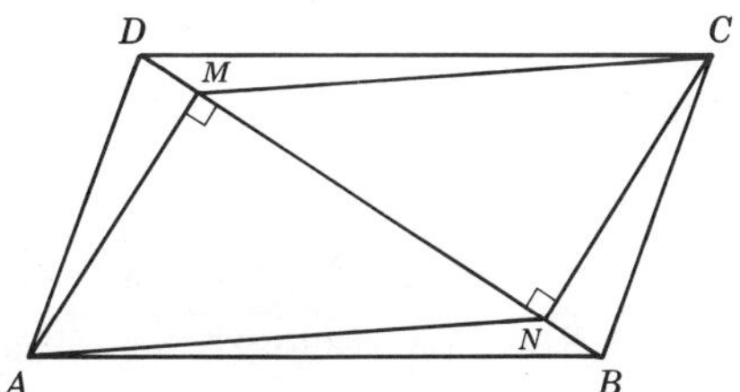

Definition 3.13 A *rectangle* is a parallelogram with one right angle.

Theorem 3.31

All angles of a rectangle are right angles.

The proof of this theorem is left as an exercise.

Theorem 3.32

The diagonals of a rectangle are equal.

Given: $ABCD$ is a rectangle

Prove: $AC = DB$.

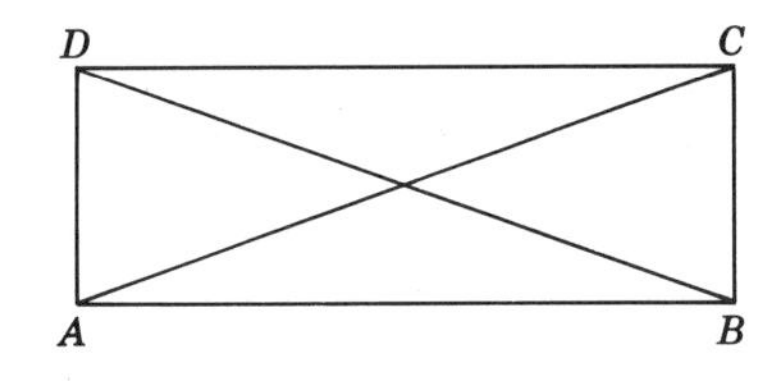

Proof:

STATEMENTS	REASONS
1. $ABCD$ is a rectangle.	1. Given.
2. $AB = DC$.	2. Since a rectangle is a parallelogram, its opposite sides are equal.
3. $\angle BAD = \angle CDA$.	3. Why?
4. $AD = DA$.	4. Why?
5. $\triangle BAD \cong \triangle CDA$.	5. SAS = SAS
6. $AC = DB$.	6. cpcte.

Theorem 3.33

If the diagonals of a parallelogram are equal, the parallelogram is a rectangle.

Given: $AC = BD$, $ABCD$ is a parallelogram.

Prove: $ABCD$ is a rectangle.

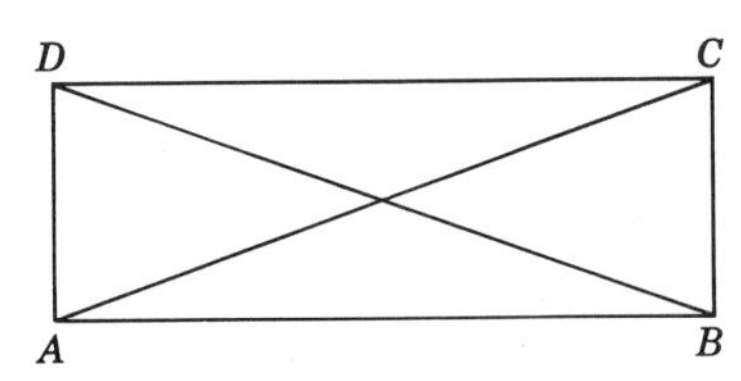

Proof:

STATEMENTS	REASONS
1. $ABCD$ is a parallelogram.	1. Given.
2. $DC = AB$.	2. Opposite sides of a parallelogram are equal.

STATEMENTS	REASONS
3. $AD = DA$.	3. Why?
4. $AC = DB$.	4. Why?
5. $\triangle ABD \cong \triangle DCA$.	5. Why?
6. $\angle BAD = \angle CDA$.	6. cpcte.
7. $\angle BAD$ and $\angle CDA$ are supplementary.	7. Why?
8. $\angle BAD = \angle CDA = 90°$.	8. If two angles are supplementary and equal, they are right angles.
9. $ABCD$ is a rectangle.	9. If a parallelogram has a right angle, it is a rectangle.

Definition 3.14 A *rhombus* is a parallelogram with two adjacent sides equal.

Theorem 3.34

All sides of a rhombus are equal.

The proof of this theorem is left as an exercise.

Theorem 3.35

The diagonals of a rhombus are perpendicular.

Given: $ABCD$ is a rhombus.

Prove: $AC \perp BD$.

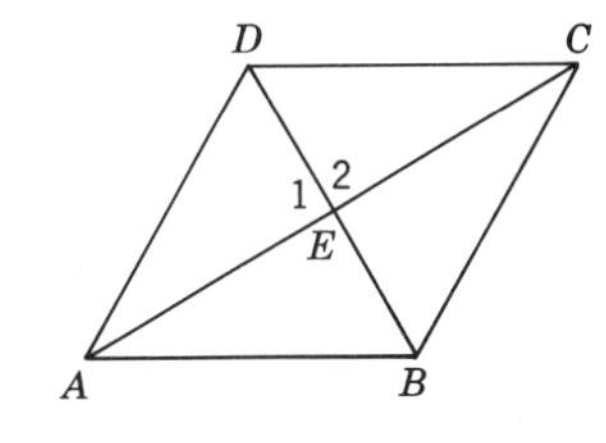

Proof:

STATEMENTS	REASONS
1. $ABCD$ is a rhombus.	1. Given.
2. $AE = CE$.	2. Since a rhombus is a parallelogram, its diagonals bisect each other.
3. $DE = DE$.	3. Why?
4. $AD = CD$.	4. Adjacent sides of a rhombus are equal.

STATEMENTS	REASONS
5. $\triangle AED \cong \triangle CED$.	5. SSS = SSS.
6. $\angle 1 = \angle 2$.	6. Why?
7. $\angle 1$ adjacent $\angle 2$.	7. Why?
8. $AC \perp BD$.	8. If two lines meet and form equal adjacent angles, they are perpendicular.

Theorem 3.36

If the diagonals of a parallelogram are perpendicular, the parallelogram is a rhombus.

Given: $BD \perp AC$; $ABCD$ is a parallelogram.

Prove: $ABCD$ is a rhombus.

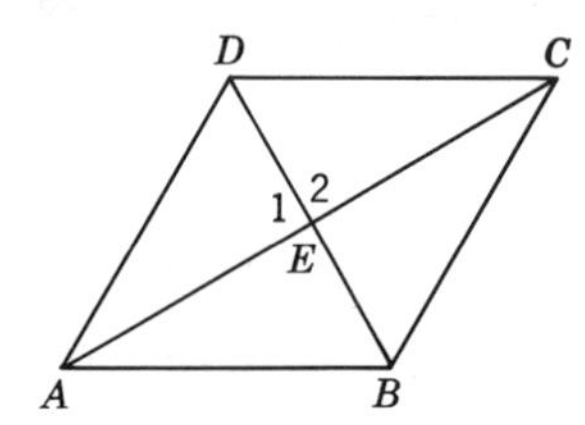

Proof:

STATEMENTS	REASONS
1. $BD \perp AC$.	1. Given.
2. $\angle 1 = \angle 2$.	2. Perpendicular lines meet and form equal adjacent angles.
3. $DE = DE$.	3. Why?
4. $ABCD$ is a parallelogram.	4. Given.
5. $AE = CE$.	5. Why?
6. $\triangle AED \cong \triangle CED$.	6. SAS = SAS.
7. $AD = CD$.	7. Why?
8. $ABCD$ is a rhombus.	8. Why?

Theorem 3.37

The diagonals of a rhombus bisect the angles of the rhombus.

The proof of Theorem 3.37 is left as an exercise.

Definition 3.15 A *square* is a rhombus with a right angle.

Theorem 3.38

If a line joins the midpoints of two sides of a triangle, that line is parallel to and equal to one-half of the third side.

Given: $\triangle ABC$ with D and E midpoints of AC and BC, respectively.

Prove: $DE \parallel AB$; $DE = \frac{1}{2}AB$.

Construction. Continue line segment DE through point E to point M so that $DE = ME$.

Proof:

STATEMENTS	REASONS
1. $\triangle ABC$ with E midpoint of BC; D midpoint of AC.	1. Given.
2. $CE = BE$.	2. A midpoint divides a line into two equal parts.
3. $\angle 1 = \angle 2$.	3. Why?
4. $DE = ME$.	4. By construction.
5. $\triangle DEC \cong \triangle MEB$.	5. Why?
6. $MB = DC$.	6. cpcte.
7. $DC = DA$.	7. Same as reason number 2.
8. $MB = DA$.	8. Quantities equal to the same quantity are equal to each other.
9. $\angle 3 = \angle 4$.	9. cpcte.
10. $MB \parallel DA$.	10. Why?
11. $ABMD$ is a parallelogram.	11. If opposite sides of a quadrilateral are both equal and parallel, the quadrilateral is a parallelogram.
12. $DE \parallel AB$.	12. Why?
13. $DM = AB$.	13. Why?
14. $DE = \frac{1}{2}DM$.	14. E is the midpoint of DM and divides DM into two equal parts each of which is one-half the whole.
15. $DE = \frac{1}{2}AB$.	15. Why?

EXERCISES

1. Prove that a square is a rectangle.
2. Prove Theorem 3.31.
3. Prove Theorem 3.34.
4. Prove Theorem 3.37.
5. *Prove:* a rectangle with two adjacent sides equal is a square.
6. AE bisects $\angle A$, BE bisects $\angle B$, $AD = 3$, $DF = 8$, and $DF \parallel AB$. Find the length of BF.

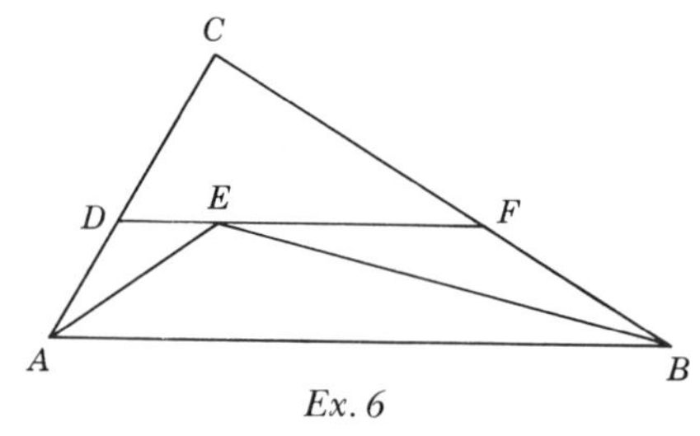

Ex. 6

*7. $ABCD$ is a quadrilateral with E, F, G and H midpoints of their respective sides. Prove $EFGH$ is a parallelogram.

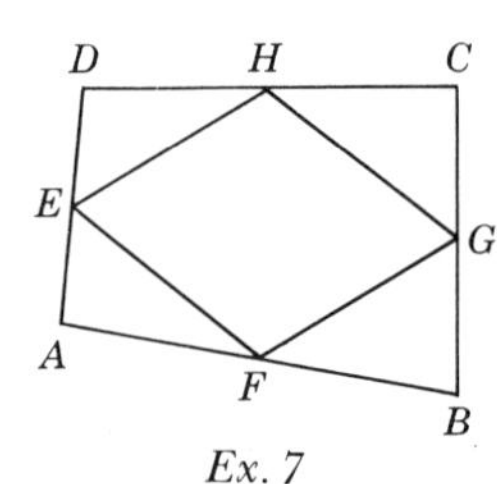

Ex. 7

8. *Prove:* the bisectors of two consecutive angles of a parallelogram are perpendicular.
9. *Prove:* the line segments joining the midpoints of the sides of an equilateral triangle form another equilateral triangle.
10. *Prove:* if two nonadjacent angles of a quadrilateral are right angles, then the bisectors of the other two angles either coincide or are parallel.
11. *Given:* $\triangle ABC$ is equilateral L, M, and N are midpoints of sides AC, CB, and AB.

    Prove: $LMBN$ is a rhombus.

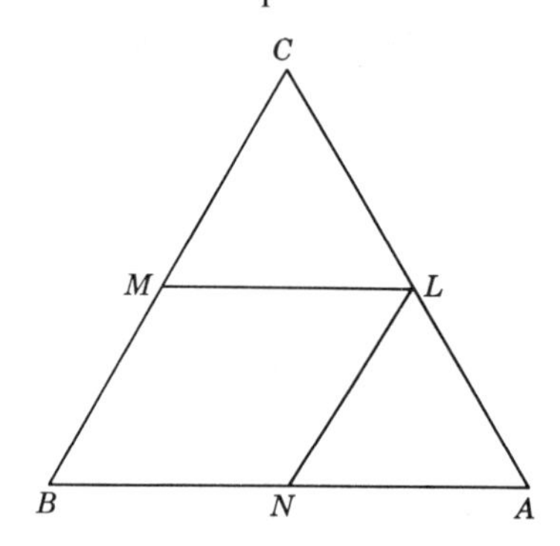

12. *Given:* $\triangle ABC$ is a right triangle L and M are midpoints of the two legs.
N is the midpoint of the hypotenuse.

Prove: $LBMN$ is a rectangle.

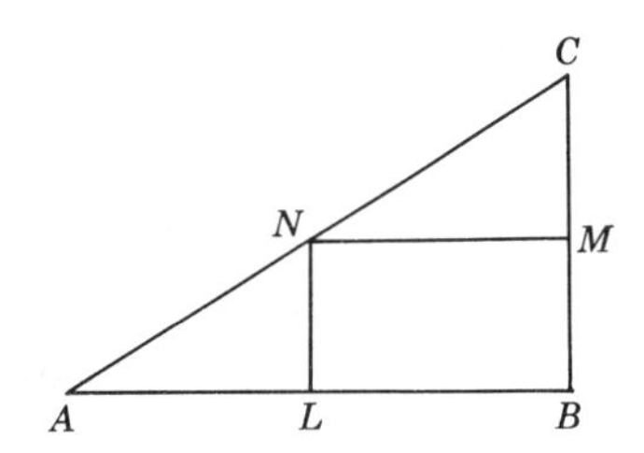

13. *Prove:* lines joining the midpoints of adjacent sides of a rhombus form a rectangle.
14. *Prove:* lines joining the midpoints of adjacent sides of a rectangle form a rhombus.

Definition 3.16 A *trapezoid* is a quadrilateral with two and only two sides parallel. The parallel sides are called *bases*.

Definition 3.17 An *isosceles trapezoid* is a trapezoid whose nonparallel sides are equal. A pair of angles including one of the parallel sides is called a pair of *base angles*.

Theorem 3.39

The base angles of an isosceles trapezoid are equal.

Given: Isosceles trapezoid $ABCD$, with $AB \parallel DC$.

Prove: $\angle A = \angle B$.

Construction. Construct perpendiculars from points D and C to line AB.

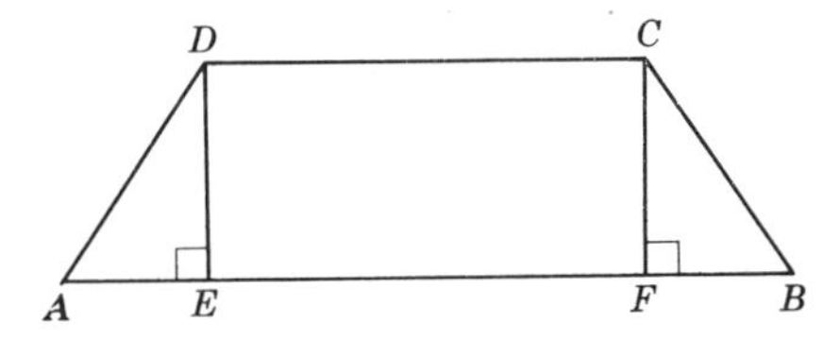

Proof:

STATEMENTS	REASONS
1. $DE \perp AB$; $CF \perp AB$, forming rt. $\triangle$s AED and BFC.	1. By construction.
2. $ABCD$ is an isosceles trapezoid.	2. Given.
3. $AD = BC$.	3. The nonparallel sides of an isosceles trapezoid are equal.
4. $DC \parallel AB$.	4. Given.
5. $DE = CF$.	5. Parallel lines are always the same distance apart.

STATEMENTS	REASONS
6. $\triangle ADE \cong \triangle BCF$.	6. Why?
7. $\angle A = \angle B$.	7. Why?

Definition 3.18 The *median of a trapezoid* is the line joining the midpoints of the non-parallel sides.

median = 1/2 times the sum of bas
OR
2 times the median = " "

Theorem 3.40

The median of a trapezoid is parallel to the bases and equal to one-half their sum.

Given: Trapezoid $ABCD$ with median EF, and $DC \parallel AB$.

Prove: $EF \parallel AB$ and $EF = \frac{1}{2}(AB + DC)$

Construction. Construct line DF and let it intersect the extension of AB at point G.

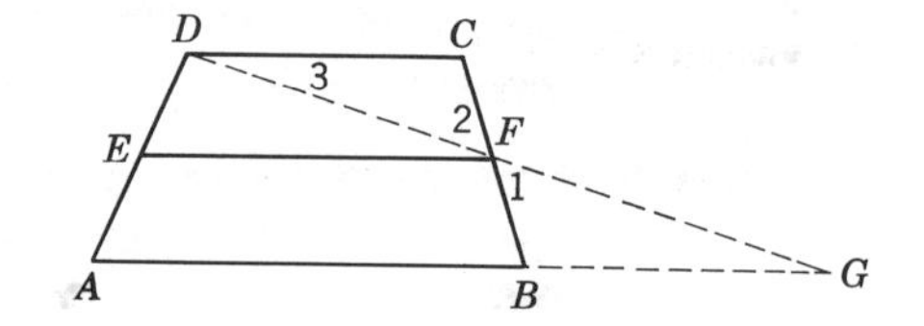

Proof:

STATEMENTS	REASONS
1. DF intersects AB at point G.	1. By construction.
2. $ABCD$ is a trapezoid with EF as median.	2. Given.
3. E midpoint of AD; F midpoint of BC.	3. Definition of "median of a trapezoid."
4. $CF = BF$, $DE = AE$.	4. Definition "midpoint."
5. $\angle 1 = \angle 2$.	5. Why?
6. $DC \parallel AB$.	6. Why?
7. $\angle 3 = \angle G$.	7. Why?
8. $\triangle DCF \cong \triangle GBF$.	8. AAS = AAS.
9. $DF = GF$.	9. cpcte.
10. $DC = GB$.	10. cpcte.
11. $EF = \frac{1}{2}AG$.	11. Why?
12. $AG = AB + GB$.	12. The whole is equal to the sum of its parts.

STATEMENTS	REASONS
13. $AG = AB + DC$.	13. Substitution.
14. $EF = \frac{1}{2}(AB + DC)$.	14. Why?
15. $EF \parallel AB$.	15. Why?

Theorem 3.41

If three or more parallel lines cut off equal segments on one transversal, they cut off equal segments on all transversals.

Given: $l_1 \parallel l_2 \parallel l_3$.
$AB = BC$.

Prove: $DE = EF$.

Construction. Construct lines parallel to AC passing through points D and E.

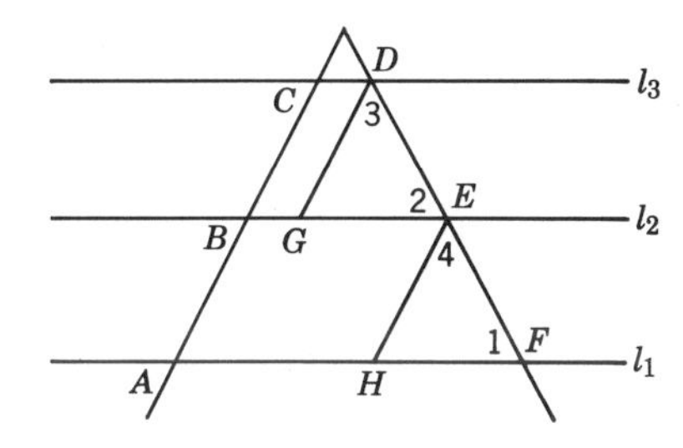

Proof:

STATEMENTS	REASONS
1. $l_1 \parallel l_2 \parallel l_3$; $AB = BC$.	1. Given.
2. $GD \parallel BC$; $HE \parallel AB$.	2. By construction.
3. $BCDG$ and $ABEH$ are parallelograms.	3. Why?
4. $BC = GD$; $AB = HE$.	4. Opposite sides of a parallelogram are equal.
5. $GD = HE$.	5. Why?
6. $GD \parallel HE$.	6. Two lines parallel to a third line are parallel to each other.
7. $\angle 1 = \angle 2$.	7. Why?
8. $\angle 3 = \angle 4$.	8. Why?
9. $\triangle GED \cong \triangle HFE$.	9. AAS = AAS.
10. $DE = EF$.	10. cpcte.

This theorem provides us with a method for dividing a line segment into any number of equal parts. Suppose that we wish to divide line segment AB into five equal parts. We first draw any ray, AC, emanating from A. We fix our compass at any convenient setting, and copy five equal segments on the ray AC, starting at the initial point, A. Thus, $AP = PQ = QR = RS = ST$. Draw line TB, and then construct other lines, parallel to TB, passing through P, Q, R, and S. The intersection of these lines with the segment AB divides AB into five equal parts as shown in Fig. 3-6.

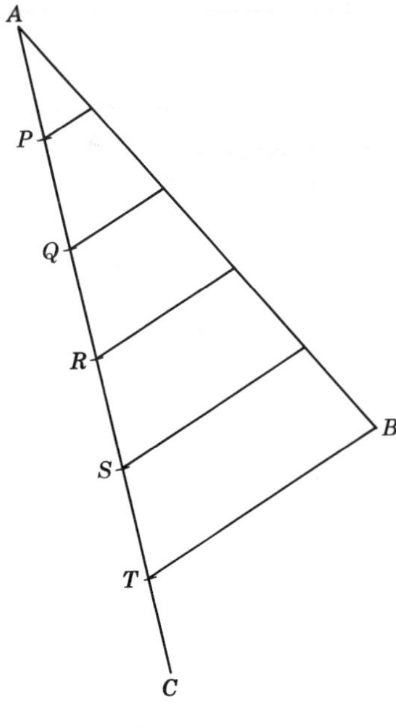

Fig. 3.6

EXERCISES

1. $ABCD$ is a trapezoid with EF as median. Find the measure of EF.
2. *Prove:* if the base angles of a trapezoid are equal, the trapezoid is an isosceles trapezoid.
3. Draw a line segment about 3 inches long. With a compass and a straightedge divide the line segment into seven equal parts.
4. $ABCD$ is a trapezoid with EF as median. Find the measure of AB.

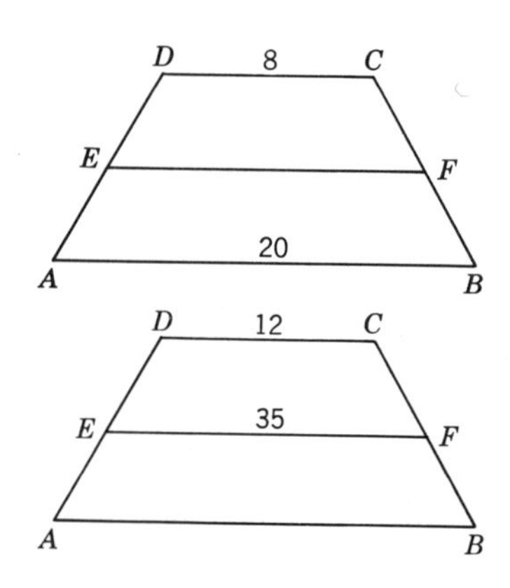

Fig. 3.7 Felix Klein was a German mathematician who first developed criteria for classifying the different geometries. In his Erlangen Program *Klein defined geometry as the study of invariants in a transformation group. This led to a unification of the ideas of Euclidean geometry, non-Euclidean geometries, projective geometry, and finally a rather new kind of geometry called topology. (Courtesy Dover Publications, Inc., publishers of Felix Klein's* Elementary Mathematics from an Advanced Standpoint).

5. *ABCD* is a parallelogram. *F*, *G*, *H*, *I* are midpoints of *DE*, *AE*, *BE*, and *CE*, respectively. Prove that *FGHI* is a parallelogram.

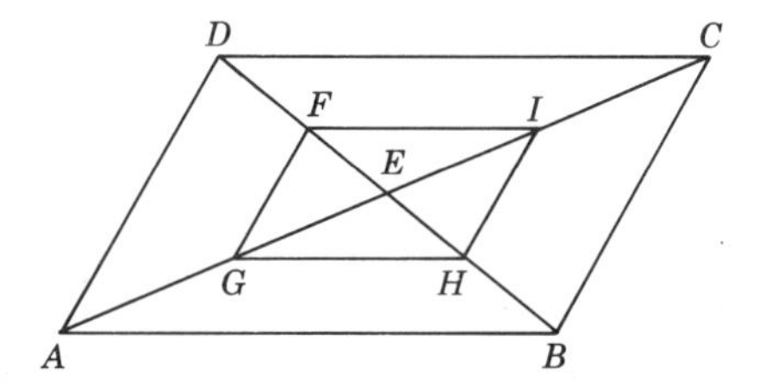

6. *Prove:* the line segments joining the midpoints of the sides of a rectangle form a rhombus.
7. *Prove:* the line segment joining the midpoints of the bases of an isosceles trapezoid is perpendicular to the bases.
8. *Prove:* the figure formed by joining the midpoints of the sides of an isosceles trapezoid is a rhombus.
9. Distinguish between the median of a triangle and the median of a trapezoid.

10. $\triangle ABC$ is equilateral. E and F are midpoints of AC and BC, respectively. G is the midpoint of AE, H is midpoint of BF. Prove that $GH = \frac{3}{4}AB$.

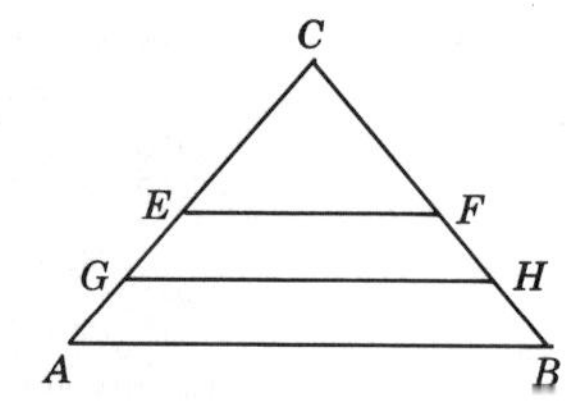

Review Test

Classify the following statements as true or false.

1. If two lines are cut by a transversal, all pairs of alternate interior angles are equal.
2. Two lines perpendicular to the same line are perpendicular to each other.
3. If a line is perpendicular to one of two perpendicular lines, it is parallel to the other.
4. Interior angles on the same side of the transversal can never be complementary.
5. The sum of the interior angles of a quadrilateral equals the sum of the exterior angles of an octagon.
6. If two angles of one triangle are equal to two angles of a second triangle, their third angles are equal.
7. The exterior angle of a triangle could be supplementary to the sum of the nonadjacent interior angles.
8. The sum of the interior angles of a seven-sided polygon equals 900°.
9. An interior angle of a polygon is greater than its corresponding exterior angle.
10. A parallelogram must have two obtuse angles.
11. A parallelogram is a special case of a trapezoid.
12. The diagonals of a rhombus are perpendicular.
13. The diagonals of a rhombus are equal.
14. The median of a trapezoid with bases of 10 inches and 22 inches has a length of 16 inches.
15. The line joining the midpoints of the bases of a trapezoid is perpendicular to the bases.
16. A triangle must have at least two acute angles.
17. An exterior angle of an equilateral triangle equals an interior angle of a regular hexagon.

18. If the diagonals of a quadrilateral are equal, then it is a rectangle.
19. Line segments joining the midpoints of opposite sides of a quadrilateral bisect each other.
20. The line segment joining the midpoints of the parallel sides of a trapezoid is perpendicular to the median of the trapezoid.

4

Areas

We now apply the results derived in the previous chapters and develop the basic formulas for finding the areas of polygons. Technically, a polygon does not have an area, although it is true that a polygon encloses an area. However, it is common to speak of the area of a polygon, and we will do so in this course.

Definition 1.20 explains what we mean by an *altitude* of a triangle. In this chapter we need similar definitions for altitudes of some other polygons.

Definition 4.1 An *altitude* of a *parallelogram* is a line segment drawn from a vertex of a parallelogram perpendicular to a nonadjacent side, or its extension. An *altitude* of a *trapezoid* is a line segment drawn from a vertex, perpendicular to the nonadjacent parallel side or its extension. The length of an altitude is called a *height*, and the side to which an altitude is drawn is called a *base*.

Postulate 4.1 The area of a rectangle is given by the formula $A = bh$ where b is the length of the base and h is the height of the rectangle.

To simplify the writing of proofs, we use the notation "$A(\triangle ABC)$" to represent the area of triangle ABC. Similarly "$A(\square ABCD)$" is used to represent the area of parallelogram $ABCD$. In general, the symbol "$A(\ )$" represents the area of the polygon that is named within the parentheses.

Postulate 4.2 Any two congruent figures have the same area.

Axiom 4.3 If a, b, and c are real numbers, $a(b+c) = ab + ac$.

Axiom 4.3 is called the *distributive property* of real numbers.

Theorem 4.1

The area of a parallelogram is given by the formula $A = bh$ where b is the length of a base and h is the corresponding height of the parallelogram.

Given: $ABCD$ is a parallelogram

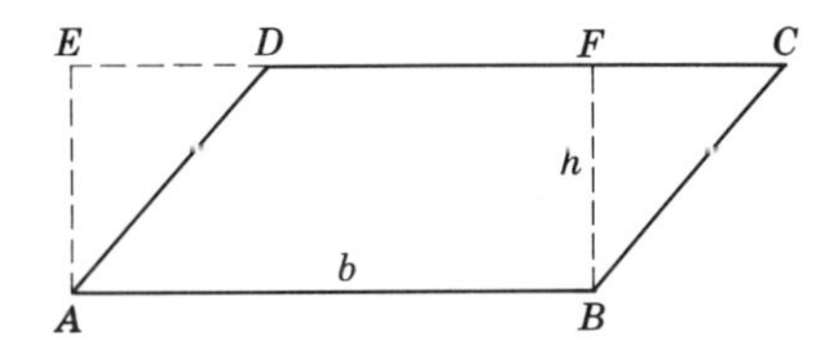

Prove: $A(\square ABCD) = bh$

Construction. Construct a perpendicular from point A to line DC extended and a perpendicular from point B to line DC.

Proof:

STATEMENTS	REASONS
1. $ABCD$ is a parallelogram.	1. Given.
2. $AD = BC$.	2. Opposite sides of a parallelogram are equal.
3. $AB \parallel DC$.	3. Why?
4. $AE \perp DC$; $BF \perp DC$.	4. By construction.
5. $AE = BF$.	5. Parallel lines are always the same distance apart.
6. $\triangle AED \cong \triangle BFC$.	6. hs = hs.
7. $A(\square ABCD) = A(\triangle BFC) + A(ABFD)$.	7. The whole is equal to the sum of its parts.
8. $A(\triangle BFC) = A(\triangle AED)$.	8. Congruent figures have equal areas.
9. $A(\square ABCD) = A(\triangle AED) + A(ABFD)$.	9. Substitution.
10. $A(ABFE) = A(\triangle AED) + A(ABFD)$.	10. The whole is equal to the sum of its parts.
11. $A(\square ABCD) = A(ABFE)$.	11. If two quantities are equal to the same quantity, they are equal to each other.
12. $AE \parallel BF$.	12. Why?
13. $ABFE$ is a rectangle.	13. Why?
14. $A(ABFE) = bh$.	14. Why?
15. $A(\square ABCD) = bh$.	15. Why?

Theorem 4.2

The area of a triangle is given by the formula $A = \frac{1}{2}bh$ where b is the length of a base and h is the corresponding height of the triangle.

Given: $\triangle ABC$.

Prove: $A(\triangle ABC) = \frac{1}{2}bh$.

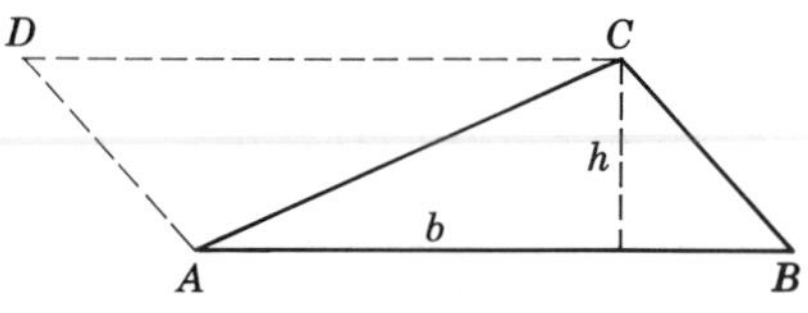

Construction. Construct a line through point C parallel to line AB and construct a line through point A parallel to line BC.

Proof:

STATEMENTS	REASONS
1. $AB \parallel DC$; $AD \parallel BC$.	1. By construction.
2. $ABCD$ is a parallelogram.	2. Why?
3. $\triangle ABC \cong \triangle CDA$.	3. Why?
4. $A(\triangle ABC) = A(\triangle CDA)$.	4. Congruent figures have equal areas.
5. $A(\square ABCD) = A(\triangle ABC) + A(\triangle CDA)$.	5. The whole is equal to the sum of its parts.
6. $A(\square ABCD) = A(\triangle ABC) + + A(\triangle ABC) = 2A(\triangle ABC)$.	6. Substitution.
7. $A(\square ABCD) = bh$.	7. Why?
8. $2A(\triangle ABC) = bh$.	8. Why?
9. $A(\triangle ABC) = \frac{1}{2}bh$.	9. Why?

Example 1

Find the area of a triangle with a base of 5 inches and a height of 8 inches.

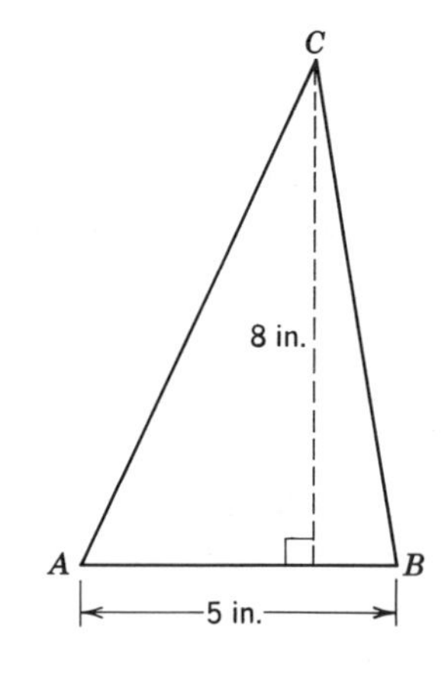

Solution: By theorem 4.2, the area A is found by the formula

$$A = \tfrac{1}{2}bh.$$

In our triangle, $b = 5$ and $h = 8$

$$A = \tfrac{1}{2}bh.$$
$$A = \tfrac{1}{2} \cdot 5 \cdot 8.$$
$$A = 20.$$

Therefore, the area of the triangle is 20 square inches.

Example 2

The area of a certain triangle is 52 square feet and the height is 13 feet. What is the measure of the base of the triangle?

Solution: Since we wish to find the length of the base, b, let us solve the equation $A = \frac{1}{2}bh$ for b. By multiplying both sides of the equation by 2, and dividing by h, we get

$$A = \tfrac{1}{2}bh.$$

$$2A = bh.$$

$$\frac{2A}{h} = b.$$

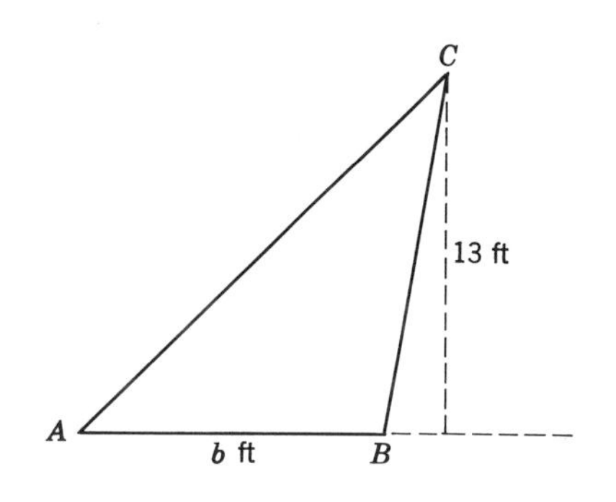

We are given that $A = 52$ and $h = 13$. Substituting into the above equation we have

$$\frac{(2)(52)}{13} = b.$$

$$8 = b.$$

Therefore, the measure of the base of the triangle is 8 feet.

A second formula for finding the area of a triangle is given in Appendix II.

Theorem 4.3

The area of a trapezoid is given by the formula $A = \frac{1}{2}h(b + b')$ where h is the height and b and b' are the lengths of the bases of the trapezoid.

Given: Trapezoid $ABCD$

Prove: $A(ABCD) = \frac{1}{2}h(b + b')$

Construction: Draw diagonal DB.

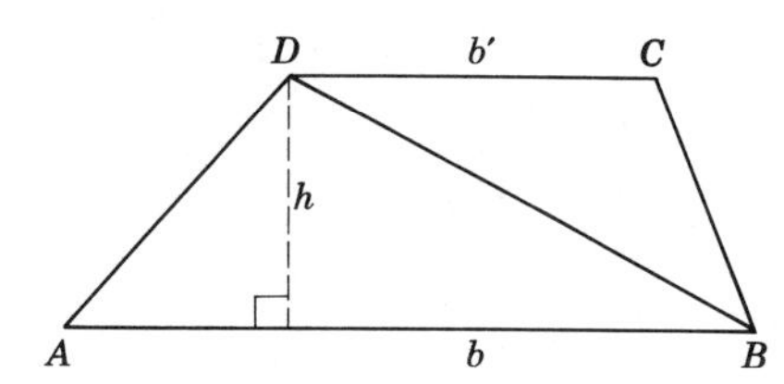

Proof:

STATEMENTS	REASONS
1. $ABCD$ is a trapezoid.	1. Given.
2. $A(ABCD) = A(\triangle ABD) + A(\triangle BCD)$.	2. The whole is equal to the sum of its parts.
3. $A(\triangle ABD) = \frac{1}{2}bh$.	3. Why?

STATEMENTS	REASONS
4. $A(\triangle BCD) = \frac{1}{2}b'h$.	4. Why?
5. $A(ABCD) = \frac{1}{2}bh + \frac{1}{2}b'h$.	5. Why?
6. $A(ABCD) = \frac{1}{2}h(b+b')$.	6. Distributive property.

Theorem 4.4

The area of a rhombus is equal to one-half the product of its diagonals.

Given: $ABCD$ is a rhombus.

Prove: $A(ABCD) = \frac{1}{2}(BD)(AC)$.

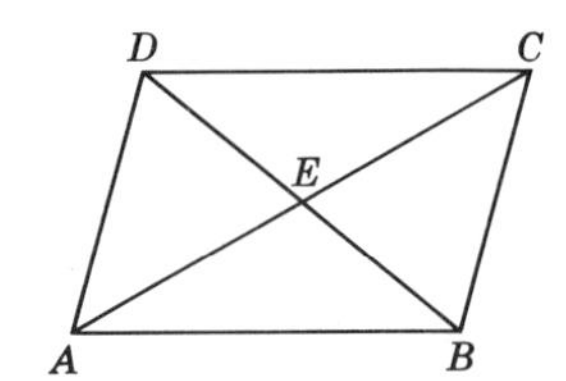

Proof:

STATEMENTS	REASONS
1. $ABCD$ is a rhombus.	1. Given.
2. $AC \perp BD$.	2. Diagonals of a rhombus are perpendicular.
3. $A(ABCD) = A(\triangle ABD) + A(\triangle CBD)$.	3. The whole is equal to the sum of its parts.
4. $A(\triangle ABD) = \frac{1}{2}(BD)(AE)$.	4. Why?
5. $A(\triangle CBD) = \frac{1}{2}(BD)(EC)$.	5. Why?
6. $A(ABCD) = \frac{1}{2}(BD)(AE) + \frac{1}{2}(BD)(EC)$.	6. Why?
7. $A(ABCD) = \frac{1}{2}(BD)(AE+EC)$.	7. Distributive property.
8. $AE + EC = AC$.	8. Why?
9. $A(ABCD) = \frac{1}{2}(BD)(AC)$.	9. Why?

Example 1

A trapezoid has bases of 6 and 10 inches, and a height of 1 foot. What is its area?

Solution: By Theorem 4.3, the area of a trapezoid is given by the formula

$$A = \frac{1}{2}h(b+b')$$

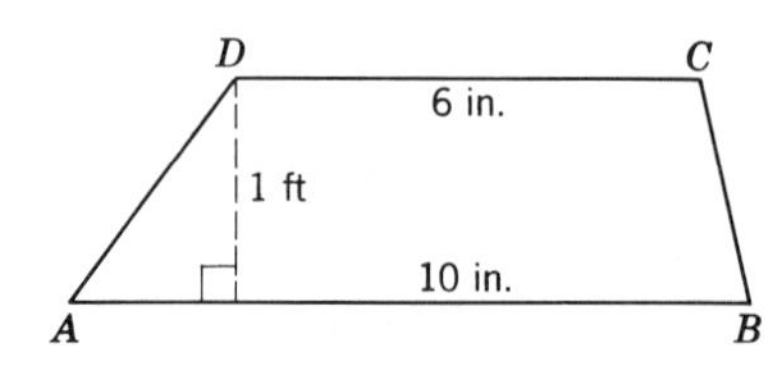

In this example, $b = 6$ and $b' = 10$. It would be incorrect to say that $h = 1$, because the height of 1 foot must be expressed in inches to be consistent with the units of the bases. Therefore, $h = 12$. Substituting in the above formula

$$A = \tfrac{1}{2}(12)(6+10)$$
$$A = \tfrac{1}{2}(12)(16)$$
$$A = 6 \cdot 16 = 96$$

Therefore the area of the trapezoid is 96 square inches.

Example 2

The wall at one end of an attic room has the shape of a trapezoid because of a slanted ceiling. The wall is 8 feet high at one end, 10 feet wide, and only 3 feet high at the other end. What is the wall's area?

Solution: The wall forms the shape of a trapezoid with bases 8 and 3, and a height 10. Substituting into the formula for the area of a trapezoid we have

$$A = \tfrac{1}{2}(10)(8+3)$$
$$A = 55$$

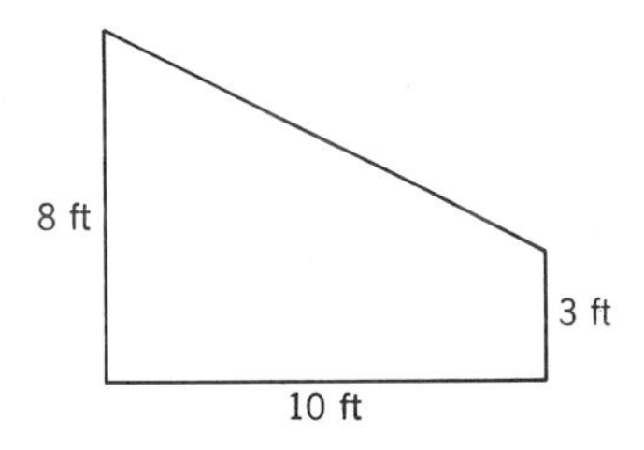

Therefore the area of the wall is 55 square feet.

A very interesting number that appears often in mathematics is *pi*, a number approximately equal to 3.14. The actual value of π is an unending decimal—3.141592653589... Calculation of π to many decimal places has occupied many mathematicians for many years. In 1873, a man named Shanks calculated π to 707 decimal places, and took 15 years to do it. Unfortunately, he made a mistake in the 528th place, and his result is wrong from that point on. With the advent of the electronic computer, π has been calculated to many thousands of decimal places.

Definition 4.2 The *circumference* of a circle is the distance around the circle. The circumference is given by the formula $C = 2\pi r$ where r is the radius of the circle.

Definition 4.2 gives us a way to evaluate π. If $C = 2\pi r$, then $\pi = C/2r$. To obtain an approximate value of π, measure the distance about some convenient circle such as the cross section of a large tin can. How can this be done? Divide this measure by twice the radius (or *diameter*) of the can. What value do you obtain?

Although the following argument is a weak proof, it is presented because it is interesting.

Conjecture. The area of a circle is given by the formula $A = \pi r^2$ where r is the radius of the circle.

Consider the circle O in Fig. 4.1. Divide the circle into an even number of equal pie-shaped pieces and regroup these pieces as in Fig. 4.2 with half of the pieces pointing up and the other half pointing down. Fig. 4.2 vaguely

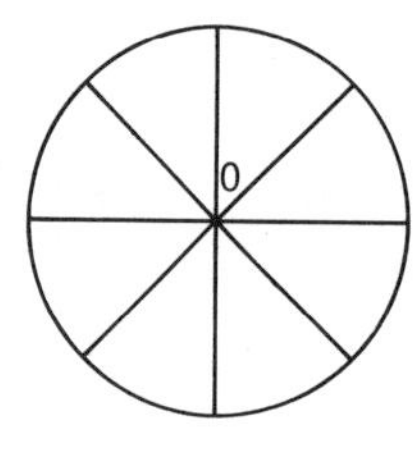

Fig. 4.1

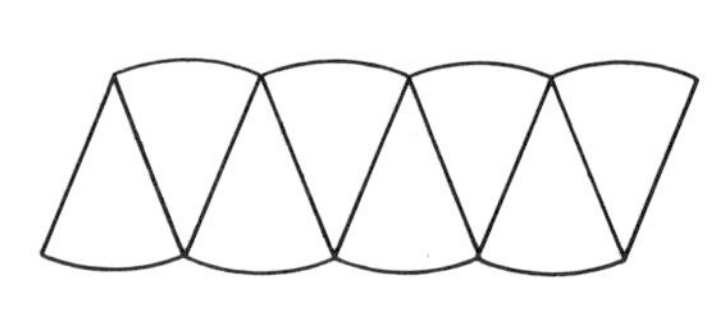

Fig. 4.2

resembles a parallelogram with base of approximately one-half of the circumference of the circle and a height approximately equal to the radius of the circle. As the number of pie-shaped pieces increases, Fig. 4.2 will look much more like a parallelogram, and the base will be much closer to one-half the circumference of the circle and the height will be much closer to the measure of a radius of the circle. We now use the formula for the area of a parallelogram

$$A = bh$$

If we let the number of pie-shaped pieces become extremely large, b becomes very close to $\frac{1}{2}C$, and h becomes very close to r.

Therefore,

$$A = bh = \tfrac{1}{2}Cr = \tfrac{1}{2}2\pi r \cdot r = \pi r^2$$

Since the areas of the pie-shaped pieces do not change when we rearrange them, the area of a circle is given by the formula

$$A = \pi r^2.$$

EXERCISES

1. Find the areas of the figures depicted below:

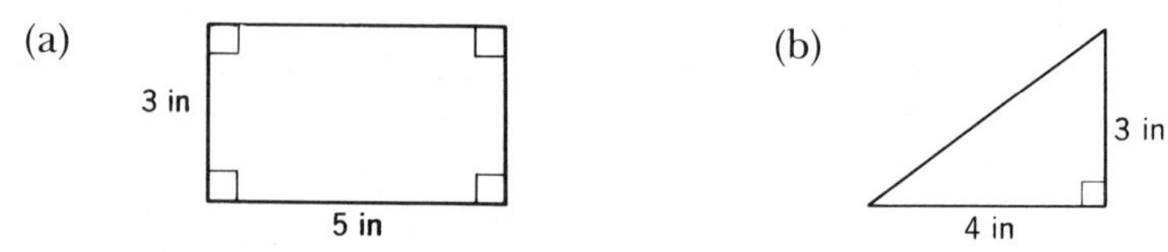

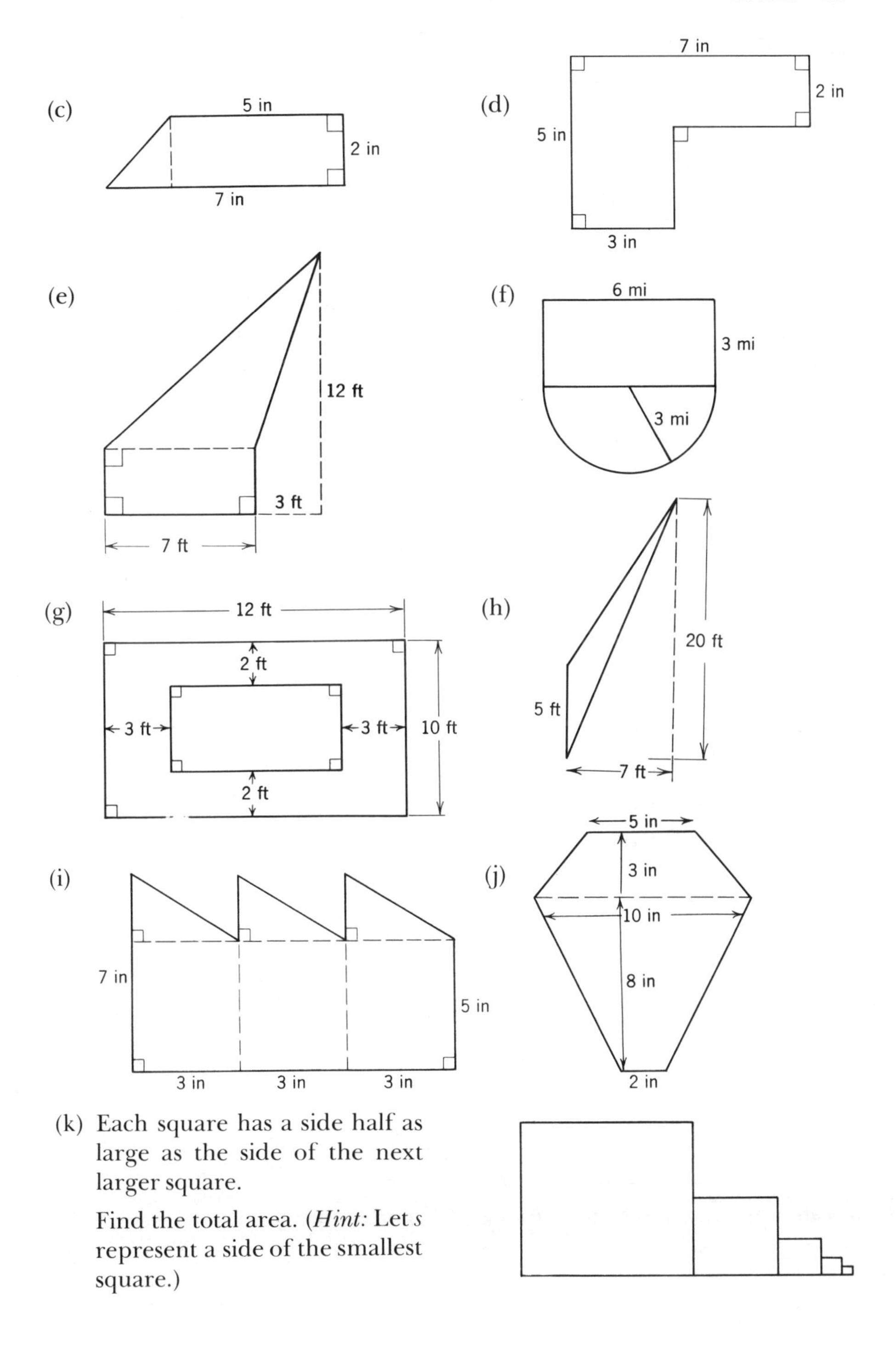

(k) Each square has a side half as large as the side of the next larger square.

Find the total area. (*Hint:* Let s represent a side of the smallest square.)

2. *Given:* ▱$ABCD$
   Point E between B and C.

   Show that the area $(AED) = \frac{1}{2}$ area (▱$ABCD$)

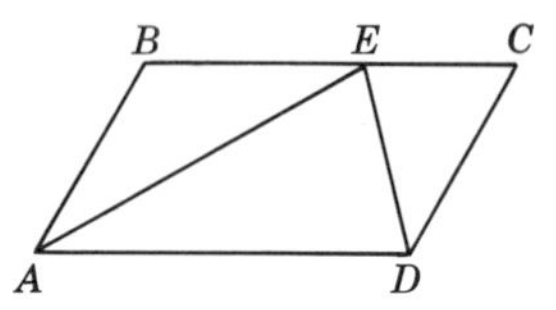

3. A rhombus has diagonals of 16 and 18 inches. Find its area.
4. A square has a diagonal of 7 inches. What is its area?
5. A square has a diagonal of $\sqrt{8}$ inches. What is the length of a side?
6. How many altitudes does a parallelogram have? How many heights? How many altitudes does a trapezoid have? How many heights?
7. *Prove:* the area of a trapezoid is given by $A = mh$, where h is the height, and m is the length of the median.
8. *Prove:* the area of a right triangle is one-half the product of lengths of the legs.

The following theorem is very important and useful. It has ramifications in many branches of mathematics as well as in many of the sciences. It was proved by Pythagoras, a Greek who lived about 500 B.C., some 200 years before Euclid, and is known as the Pythagorean theorem.

Theorem 4.5

In a right triangle, the square of the hypotenuse is equal to the sum of the squares of the other two sides.

Given: $\triangle ABC$ is a right triangle with $\angle C = 90°$.

Prove: $a^2 + b^2 = c^2$

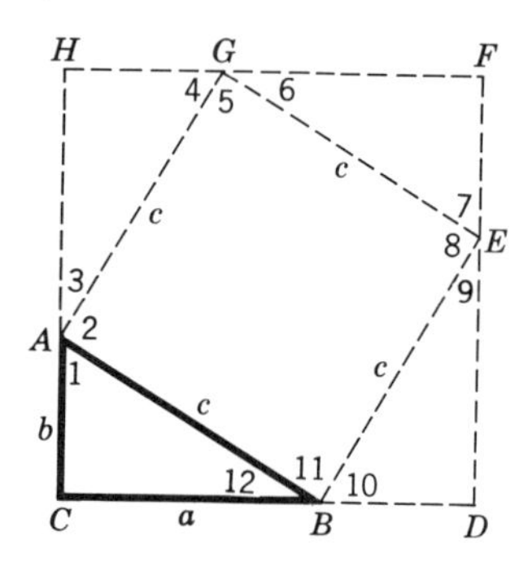

Proof: Construct a square $AGEB$ on the hypotenuse, AB, of the right triangle ABC. Through point G, construct a line parallel to line CB. Through point E, construct a line parallel to line AC. By this construction, angles 2, 5, 8, and 11 are all right angles, and sides AG, GE, EB, and BA are equal.

Fig. 4.3 Pythagoras (about 584–495 B.C.). Pythagoras was a Greek philosopher and mathematician, and the leader of a secret society that believed in numerology. Although Pythagoras is most famous for the theorem that bears his name, he is sometimes called the "father of music." The Pythagoreans are credited with the discovery of the fundamentals of musical harmony. They realized the mathematical relationship between the length of a vibrating string and its pitch. They are also credited with the discovery of irrational numbers. This representation is from the Cathedral of Chartres, France. (New York Public Library Picture Collection.)

Since $\angle C$ is a right angle, and $HF \parallel CD$, and $HC \parallel FD$, we know that $CHFD$ is a rectangle, and therefore angles H, F, and D are right angles. We now show that the four triangles in the drawing are congruent.

Angle GAC (equal to $\angle 1 + \angle 2$) is an exterior angle of $\triangle AGH$, and is therefore equal to the sum of the nonadjacent interior angles, $\angle 4$ and $\angle H$:

$$\angle 1 + \angle 2 = \angle 4 + \angle H$$

But $\angle 2$ is 90°, and $\angle H$ is 90°, so that

$$\angle 1 + 90° = \angle 4 + 90°$$

or

$$\angle 1 = \angle 4.$$

Since BA and AG are equal, we know by ha = ha that $\triangle ABC \cong \triangle GAH$. In a similar manner we can show the other triangles congruent as well.

Since each side of the large square is $(a+b)$ (can you explain why?), we know its area to be $(a+b)^2$, or $a^2+2ab+b^2$. The small square has side of length c, so its area is c^2. The area of one triangle is $\frac{1}{2}ab$, and since the triangles are all congruent, their total area is four times this, or $2ab$.

Since the whole is equal to the sum of its parts, we know that the area of the large square, $CHFD$, equals the sum of the areas of the small square, $AGEB$, and the four triangles. Therefore

$$a^2+2ab+b^2 = 2ab+c^2$$

or

$$a^2+b^2 = c^2.$$

Example 1

The legs of a right triangle are 3 feet and 4 feet in length. What is the length of the hypotenuse of the triangle?

Solution: By the Pythagorean theorem, the sum of the squares of the two legs, a and b, of a right triangle is equal to the square of the hypotenuse, c:

$$a^2+b^2 = c^2$$

We are given the two legs: $a = 3$ and $b = 4$. Substituting in the above formula,

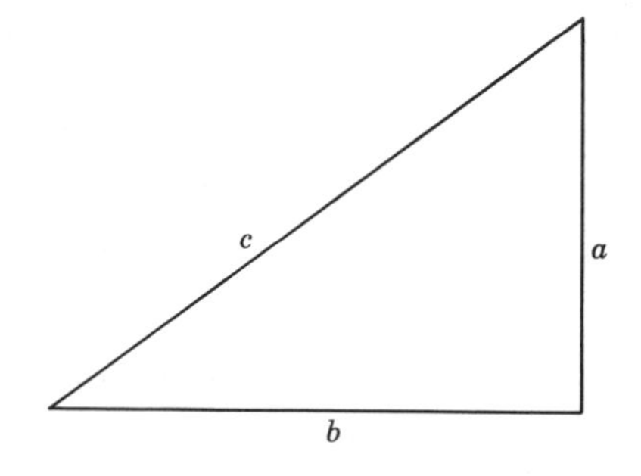

$$a^2+b^2 = c^2$$
$$3^2+4^2 = c^2$$
$$9+16 = c^2$$
$$25 = c^2$$
$$\sqrt{25} = c$$
$$5 = c$$

Therefore the length of the hypotenuse is 5 feet.

Example 2

The legs of a certain right triangle are equal and the hypotenuse is $\sqrt{8}$. What is the length of a leg of the triangle?

Solution: Let x represent the length of each leg. By the Pythagorean theorem,

$$x^2 + x^2 = (\sqrt{8})^2$$
$$2x^2 = 8$$
$$x^2 = 4$$
$$x = 2$$

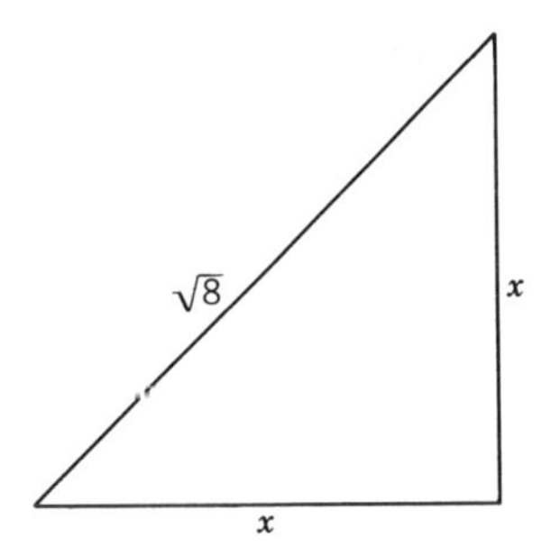

Therefore each leg is 2 units in length.

We now prove the converse of the Pythagorean theorem.

Theorem 4.6

If a triangle has sides of length *a*, *b*, and *c* and $c^2 = a^2 + b^2$, then the triangle is a right triangle.

Given: $\triangle ABC$ with $c^2 = a^2 + b^2$.

Prove: $\triangle ABC$ is a right triangle.

Plan: Construct right triangle EFG, with $e = a$, $f = b$, and $\angle 2 = 90°$. We will probe $\triangle ABC \cong \triangle EFG$ and thereby show that $\angle 1 = \angle 2 = 90°$.

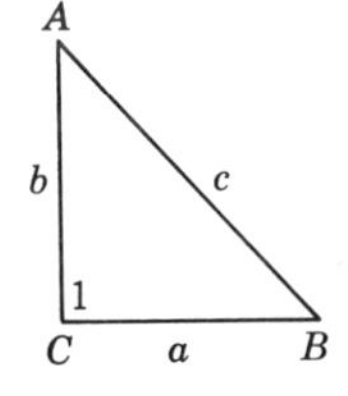

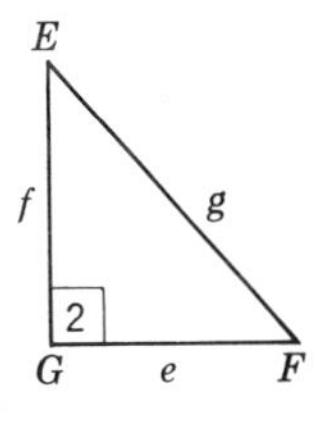

Proof:

STATEMENTS	REASONS
1. $\triangle ABC$ with $c^2 = a^2 + b^2$.	1. Given.
2. $\triangle EFG$ with $e = a, f = b$ and $\angle 2 = 90°$.	2. By construction.
3. $g^2 = e^2 + f^2$.	3. Pythagorean theorem.
4. $e^2 = a^2; f^2 = b^2$.	4. Why?
5. $g^2 = a^2 + b^2$.	5. Substitution.
6. $g^2 = c^2$.	6. Why?
7. $g = c$.	7. Property of algebra. (Since g and c represent lengths, they must both be positive.)
8. $\triangle ABC \cong \triangle EFG$.	8. SSS = SSS.

STATEMENTS	REASONS
9. $\angle 1 = \angle 2$.	9. cpcte.
10. $\angle 1 = 90°$	10. Substitution.
11. $\triangle ABC$ is a right triangle.	11. A triangle with a right angle is a right triangle.

Example 1

A triangle has sides that measure 20, 21 and 29 inches. Determine if the triangle is a right triangle.

Solution: By Theorem 4.6, we know we have a right triangle if we can show that the square of one side is equal to the sum of the squares of the other two sides of the triangle. Let $a = 20$, $b = 21$, and $c = 29$ and compute their squares,

$$a^2 = 400$$
$$b^2 = 441$$
$$c^2 = 841$$

We notice that $a^2 + b^2 = c^2$. Therefore the triangle is a right triangle.

Example 2

Is a triangle with sides of 3, 7 and 11 inches a right triangle?

Solution: Let $a = 3$, $b = 7$, and $c = 11$. We compute the squares:

$$a^2 = 9$$
$$b^2 = 49$$
$$c^2 = 121$$

Is the sum of any two of these squares equal to the third? The answer is; "no"; the given triangle is not a right triangle.

Theorem 4.7

In a 30°–60° right triangle, the hypotenuse is twice the length of the side opposite the 30° angle. The side opposite the 60° angle is equal to the length of the side opposite the 30° angle multiplied by $\sqrt{3}$.

Given: Right triangle ABC with $\angle A =$ 60° and $\angle B = 30°$.

Prove: $AB = 2AC$; $BC = AC\sqrt{3}$.

Construction. Construct a 30° angle, $\angle CBD$, where point D lies on the extension of side AC.

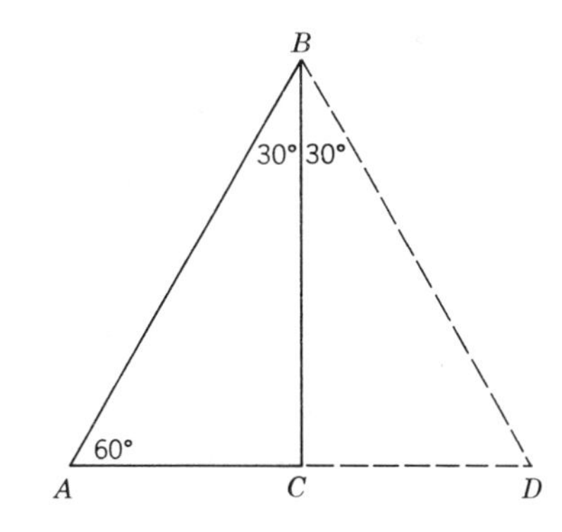

Proof:

STATEMENTS	REASONS
1. $\triangle ABC$; $\angle A = 60°$, $\angle B = 30°$.	1. Given.
2. $\triangle DBC$; $\angle CBD = 30°$.	2. By construction.
3. $\angle D = 60°$.	3. The sum of the angles of a triangle equals 180°.
4. $\triangle ABD$ is equiangular.	4. Why?
5. $\triangle ABD$ is equilateral.	5. Why?
6. $AC = CD$.	6. Why?
7. $AD = AC + CD$.	7. The whole is equal to the sum of its parts.
8. $AD = AC + AC = 2AC$.	8. Why?
9. $AD = AB$.	9. Why?
10. $AB = 2AC$.	10. Substitution.
11. $\triangle ABC$ is a right triangle.	11. Why?
12. $(AB)^2 = (AC)^2 + (BC)^2$.	12. Why?
13. $4(AC)^2 = (AC)^2 + (BC)^2$.	13. Substitution.
14. $3(AC)^2 = (BC)^2$.	14. Why?
15. $(AC)\sqrt{3} = (BC)$.	15. Why?

Theorem 4.8

In an isosceles 45° right triangle, the hypotenuse is equal to the length of one of its arms multiplied by $\sqrt{2}$.

The proof of this theorem is left as an exercise.

Example 1

An equilateral triangle has sides of 8 inches. What is its height?

Solution. Draw the bisector of $\angle C$. $\triangle ADC$ is a 30°–60° right triangle, with $\angle A = 60°$. Why? By Theorem 4.7, the side opposite the 30° angle is half as long as the hypotenuse. Therefore, $AD = 4$ inches. The height of the equilateral triangle is CD, which by Theorem 4.7 is $\sqrt{3}$ times AD. The height is $4\sqrt{3}$ inches.

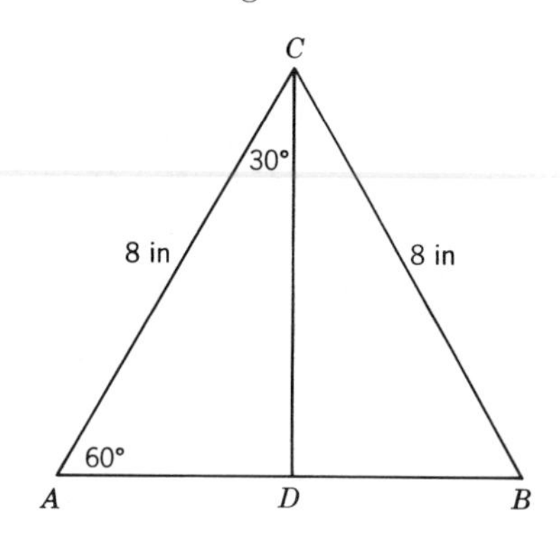

Example 2

A square has sides of $\sqrt{8}$ feet. How long is its diagonal?

Solution. Draw diagonal AC. $\triangle ABC$ is an isosceles right triangle. Why? By Theorem 4.8 the hypotenuse AC is equal to the length of an arm multiplied by $\sqrt{2}$. Therefore $AC = \sqrt{8} \cdot \sqrt{2} = \sqrt{16} = 4$ feet.

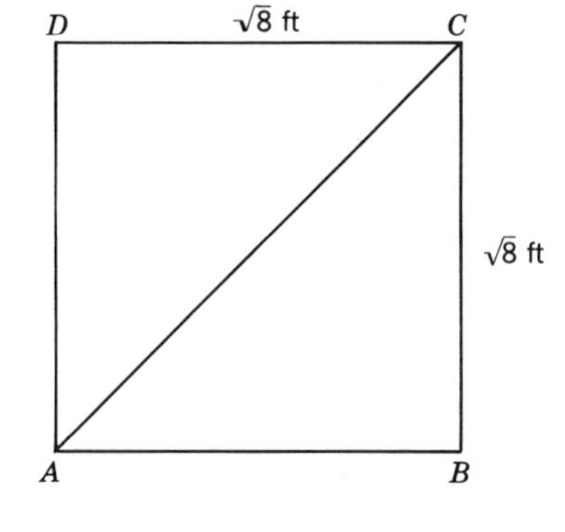

Alternate solution: $\triangle ADC$ is a right triangle with arms of $\sqrt{8}$. We may apply the Pythagorean theorem:

$$a^2 + b^2 = c^2$$
$$(\sqrt{8})^2 + (\sqrt{8})^2 = c^2$$
$$8 + 8 = c^2$$
$$16 = c^2$$
$$4 = c$$

Therefore the diagonal is 4 feet long.

EXERCISES

1. Prove Theorem 4.8.
2. The sides of a square measure 2 inches. How long is a diagonal?

3. The hypotenuse of a 30°–60° right triangle is 4 inches. Find the lengths of the legs.

4. Find the area of the triangle in problem 3.

5. *ABC* is an equilateral triangle. Find its perimeter.

6. Find the length of the altitude of equilateral triangle *ABC*.

7. Find the area of equilateral triangle *ABC*.

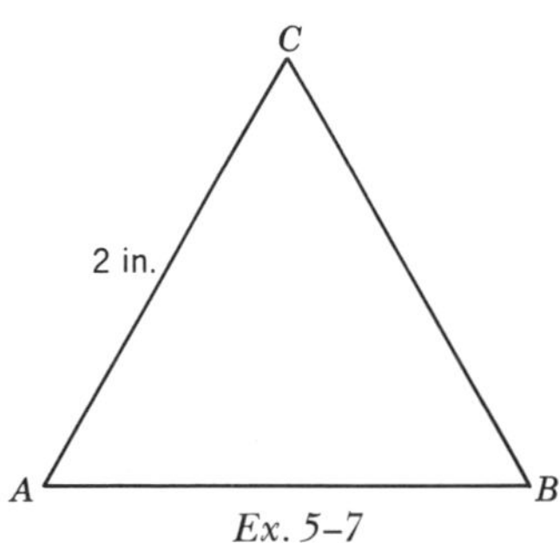

Ex. 5–7

8. Devise a construction technique that multiplies the length of a given line segment by $\sqrt{2}$.

9. Devise a construction technique that divides the length of a given line segment by $\sqrt{3}$.

10. A cube has edges of one inch. How long is diagonal *AB*?

11. The altitude of an equilateral triangle is 8 inches long. Find the length of a side of the triangle.

12. A rectangle has sides of 4 and 9 feet. Find the side of a square with same area.

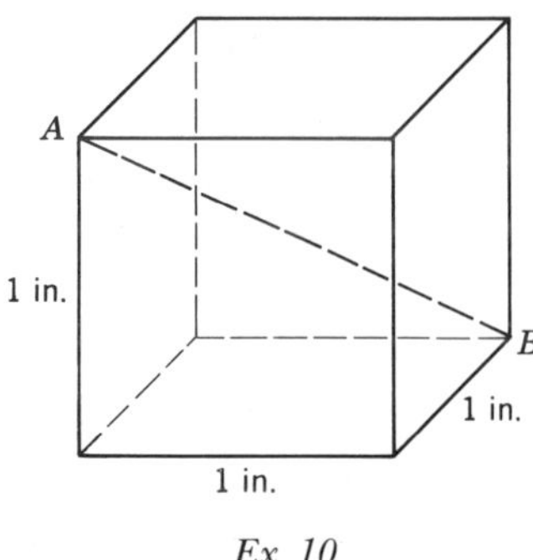

Ex. 10

13. A ladder that is 13 feet long is placed against a building. The foot of the ladder is 5 feet from the base of the building. How far above the ground does the ladder touch the building?

14. A television tower is 150 feet tall and is secured by three guy wires fastened at the top of the tower and to hooks 15 feet from the base of the tower. If the tower is installed on a flat roof, how much guy wire is used?

15. Is a triangle with sides of 25, 60, and 65 inches a right triangle?

16. Figure 4.4 illustrates a cube with edges of one foot. Points *S* and *F* are the midpoints of their respective edges. A spider at point *S* walks via the shortest path to a fly at point *F*. How far does the spider walk?

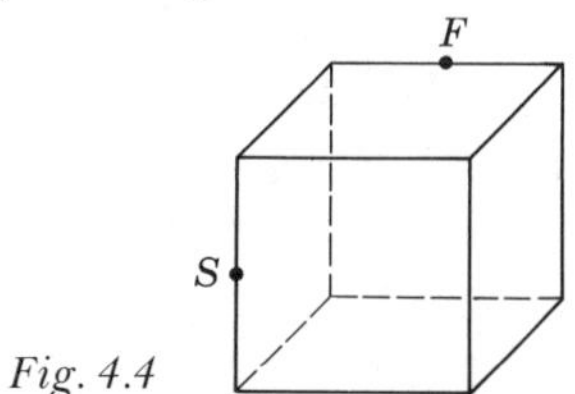

Fig. 4.4

Review Test

Classify the following statements as true or false.

1. If two figures are congruent, they have the same area.
2. If two figures have the same area, then they are congruent.
3. A triangle with sides of 3, 4, and 5 inches is a right triangle.
4. A triangle with sides of 5, 24, and 25 inches is a right triangle.
5. A triangle with sides of 3, 5, and 7 inches is a right triangle.
6. The distance around a circle is called its perimeter.
7. The area of a parallelogram is equal to the product of its diagonals.
8. A rectangle has sides of 6 and 8 inches. A square with the same area has a side of 7 inches.
9. A triangle has three altitudes.
10. The area of a circle is given by the formula: $A = \pi r^2$.
11. The side opposite the 30° angle in a right triangle with a hypotenuse 6 inches long is 12 inches long.
12. If one side of a triangle is $\sqrt{2}$ times another side, the triangle is a right triangle.
13. The area of an equilateral triangle with a height of 6 inches is $12\sqrt{3}$ square inches.
14. The area of a parallelogram with sides of 10 inches and 14 inches is 140 square inches.
15. If two trapezoids have equal heights and equal medians, they have equal areas.
16. A rectangle is a parallelogram.
17. The area of a trapezoid with bases of 12 inches and 18 inches is $15h$ where h is the height of the trapezoid.
18. The area and circumference of a circle with a radius of 2 inches have the same number of units in their measure.
19. A square with sides of 3 inches has the same area as a circle with a radius of 3 inches.
20. A square has a diagonal of $\sqrt{13}$ inches. The area of the square is 13/2 square inches.

5

Ratio, Proportion and Similarity

In this chapter we study ratio, proportion, and similarity of triangles. We have previously studied ratio and proportion in algebra and now rely heavily on the results from that course. On page 9 and page 78 several axioms from algebra are stated. In addition to these, we assume familiarity with the basic manipulations of fractions.

Definition 5.1 A *ratio* is the comparison of two numbers by their indicated quotient. A ratio is a fraction. Of course, denominators are not equal to zero.

Definition 5.2 A *proportion* is a statement that two ratios are equal.

A proportion is usually written as $a/b = c/d$. In this chapter we assume that none of the terms a, b, c or d are zero.

Definition 5.3 In the proportion

$$\frac{a}{b} = \frac{c}{d}$$

the numbers a and d are called the *extremes* of the proportion, and the numbers b and c are called the *means* of the proportion. The single term, d, is called the *fourth proportional*.

Let us experiment with a simple example of a proportion and see if we can discover some of the properties of proportions. Since the fractions 2/6 and 3/9 are both reducible to 1/3, it is true that $2/6 = 3/9$. In this proportion, the means are 6 and 3 and the extremes are 2 and 9. We notice that the product of the means and the product of the extremes are equal; $6 \cdot 3 = 2 \cdot 9$. Experimenting with other examples will lead us to suspect that this is a general property of proportions. This property is formally proved in Theorem 5.1.

What else might we discover about the proportion $2/6 = 3/9$? We might notice that the denominator of each side can be added to the numerator, to produce another true proportion.

$$\frac{2+6}{6} = \frac{3+9}{9}, \qquad \text{or} \qquad \frac{8}{6} = \frac{12}{9}$$

because both sides reduce to 4/3. Again, more experimenting leads us to believe this may be a general property of proportions. This property is proved in Theorem 5.5.

There are many properties of proportions; we now prove several.

Theorem 5.1

In a proportion, the product of the means is equal to the product of the extremes.
(If $a/b = c/d$, then $ad = bc$.)

Given: $\frac{a}{b} = \frac{c}{d}$

Prove: $ad = bc$

Proof:

STATEMENTS	REASONS
1. $\frac{a}{b} = \frac{c}{d}$.	1. Given.
2. $bd = bd$.	2. A quantity is equal to itself.
3. $\frac{a}{b}(bd) = \frac{c}{d}(bd)$.	3. Equals multiplied by equals are equal.
4. $ad = bc$.	4. Property of fractions.

Theorem 5.2

A proportion may be written by inversion.
(If $a/b = c/d$, then $b/a = d/c$.)

Given: $\frac{a}{b} = \frac{c}{d}$

Prove: $\frac{b}{a} = \frac{d}{c}$

Proof:

STATEMENTS	REASONS
1. $\frac{a}{b} = \frac{c}{d}$.	1. Given.
2. $ad = bc$.	2. The product of the means is equal to the product of the extremes.
3. $ac = ac$.	3. Why?
4. $\frac{ad}{ac} = \frac{bc}{ac}$.	4. Why?
5. $\frac{d}{c} = \frac{b}{a}$.	5. Property of fractions.

Theorem 5.3

The means may be interchanged in any proportion. (If $a/b = c/d$, then $a/c = b/d$.)

Given: $\frac{a}{b} = \frac{c}{d}$

Prove: $\frac{a}{c} = \frac{b}{d}$

Proof:

STATEMENTS	REASONS
1. $\frac{a}{b} = \frac{c}{d}$.	1. Given
2. $ad = bc$.	2. Why?
3. $cd = cd$.	3. Why?
4. $\frac{ad}{cd} = \frac{bc}{cd}$.	4. Why?
5. $\frac{a}{c} = \frac{b}{d}$.	5. Why?

Theorem 5.4

The extremes may be interchanged in any proportion. (If $a/b = c/d$, then $d/b = c/a$.)

The proof of Theorem 5.4 is left as an exercise.

Theorem 5.5

A proportion may be written by addition.
(If $a/b = c/d$, then $(a+b)/b = (c+d)/d$.)

Given: $\frac{a}{b} = \frac{c}{d}$

Prove: $\frac{a+b}{b} = \frac{c+d}{d}$

Proof:

STATEMENTS	REASONS
1. $\frac{a}{b} = \frac{c}{d}$.	1. Given.
2. $\frac{a}{b} + 1 = \frac{c}{d} + 1$.	2. Equals added to equals are equal.
3. $\frac{a}{b} + \frac{b}{b} = \frac{c}{d} + \frac{d}{d}$.	3. Property of fractions.
4. $\frac{a+b}{b} = \frac{c+d}{d}$.	4. Addition of fractions.

Theorem 5.6

A proportion may be written by subtraction.
(If $a/b = c/d$, then $(a-b)/b = (c-d)/d$.)

The proof of this theorem is left as an exercise.

Theorem 5.7

If three terms of one proportion are equal, respectively, to three terms of a second proportion, the fourth terms are equal.

The proof of this theorem is left as an exercise.

Theorem 5.8

If $a/b = c/d = e/f$, then $\frac{a+c+e}{b+d+f} = \frac{a}{b}$

Given: $\frac{a}{b} = \frac{c}{d} = \frac{e}{f}$

Prove: $\frac{a+c+e}{b+d+f}=\frac{a}{b}$

Hint: Let $\frac{a}{b}=r$

Proof:

STATEMENTS	REASONS
1. $\frac{a}{b}=\frac{c}{d}=\frac{e}{f}=r.$	1. Given.
2. $a=br$ $c=dr.$ $e=fr.$	2. Equals multiplied by equals are equal.
3. $a+c+e=br+dr+fr.$	3. Equals added to equals are equal.
4. $a+c+e=r(b+d+f).$	4. Distributive property.
5. $r=\frac{a+c+e}{b+d+f}.$	5. Equals divided by equals are equals.
6. $\frac{a+c+e}{b+d+f}=\frac{a}{b}.$	6. Substitution.

Example 1

In the proportion $\frac{x}{2}=\frac{3}{6}$, what is the value of x?

Solution. In any proportion, the product of the means equals the product of the extremes. Therefore

$$6x=2\cdot 3$$
$$6x=6$$
$$x=1$$

Example 2

In the proportion $\frac{x}{3}=\frac{3}{x}$, what is x?

Solution: By Theorem 5.1

$$x\cdot x=3\cdot 3$$
$$x^2=9$$
$$x=3 \qquad \text{or} \qquad x=-3$$

Example 3

In the proportion $\frac{x}{2} = \frac{\frac{3}{2}+x}{5}$, what is x?

Solution: By Theorem 5.1

$$5x = 3 + 2x$$

Subtracting $2x$ from both sides of the equation,

$$3x = 3$$
$$x = 1$$

Example 4

Solve the equation $\frac{52}{x} = 13$ for x.

Solution: Since 13 can be written as the fraction, 13/1, the problem is to solve the proportion

$$\frac{52}{x} = \frac{13}{1}$$

By Theorem 5.1

$$13x = 1 \cdot 52$$
$$13x = 52$$
$$x = \frac{52}{13}$$
$$x = 4$$

Example 5

Solve the proportion $\frac{n+3}{n} = \frac{4}{3}$.

Solution. By Theorem 5.1

$$4 \cdot n = 3(n+3)$$
$$4n = 3n + 9$$
$$n = 9$$

Alternate solution: Writing the proportion by subtraction we have

$$\frac{n+3-n}{n} = \frac{4-3}{3}$$

or

$$\frac{3}{n} = \frac{1}{3}.$$

By Theorem 5.1

$$n \cdot 1 = 3 \cdot 3$$
$$n = 9$$

EXERCISES

1. Solve for n: $\dfrac{n-2}{n} = \dfrac{3}{5}$.
2. Solve for n: $\dfrac{3}{2n} = \dfrac{n/2}{3}$.
3. Solve for n: $\dfrac{4}{3} = \dfrac{n+3}{5}$.
4. Solve for n: $\dfrac{n+1}{n} = \dfrac{2}{3}$.
5. Prove Theorem 5.4.
6. Prove Theorem 5.6.
7. Verify that each of the Theorems 5.1 through 5.6 is true when applied to the proportion $2/6 = 3/9$.
8. In the proportion $2/6 = 3/9$, we might notice that $2/(6+2) = 3/(9+3)$, because $2/8$ does equal $3/12$. Is this property true for all proportions? If $a/b = c/d$, is $a/(b+a) = c/(d+c)$? If so, prove it.
9. In the proportion $a/b = c/d$, which proportion theorems would fail if $a = 0$ or $c = 0$?
10. Solve the proportion $x/2 = x/5$.
11. Solve for x: $\dfrac{2}{x} = \dfrac{5}{x}$.
12. Distinguish between a ratio and a proportion.
13. Prove Theorem 5.7.
14. Solve for x: $\dfrac{x+1}{6} = \dfrac{x-1}{4}$
15. Solve for x: $\dfrac{2}{x-3} = \dfrac{x}{2}$
16. Does $\dfrac{791}{1031} = \dfrac{2373}{3092}$?

17. Find the fourth proportional to 3, 9, and 17.

18. Solve for x: $\frac{x+2}{x^2+5x+6}=\frac{x+5}{x^2+8x+15}$.

If two triangles have the same shape, they are said to be *similar*. Since congruent triangles have the same shape they are similar triangles. However, similar triangles are not necessarily congruent triangles. We can think of one of two similar triangles as being a magnification of the other. Fig. 5.1 shows a pair of similar triangles that are not congruent.

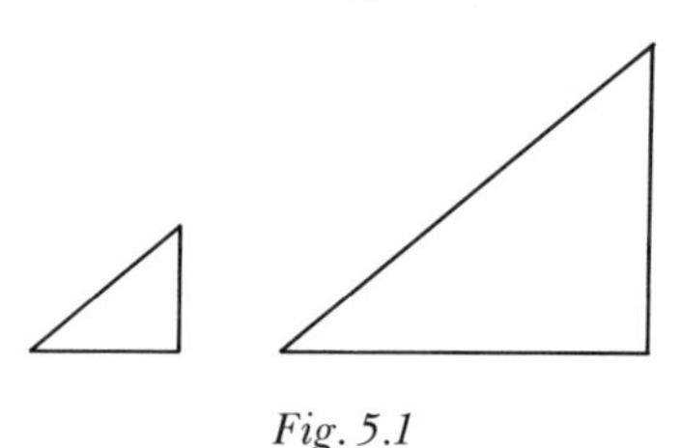

Fig. 5.1

Definition 5.4 *Two triangles are similar* if, and only if, three angles of one triangle are equal to three angles of the second triangle and all pairs of corresponding sides are in proportion. The symbol "$\sim$" means "is similar to."

Postulate 5.1 Two triangles are similar if, and only if, two angles of one triangle are equal to two angles of the other triangle.

Theorem 5.9

A line parallel to one side of a triangle divides the other two sides proportionally.

Given: $\triangle ABC$
$DE \parallel AB$.

Prove: $\frac{AD}{DC}=\frac{BE}{EC}$.

Proof:

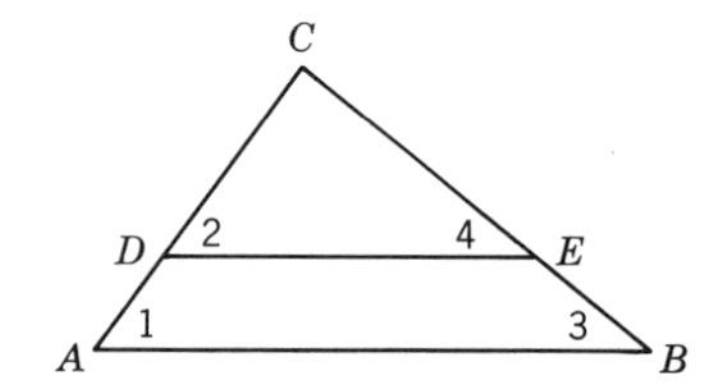

STATEMENTS	REASONS
1. $AB \parallel DE$.	1. Given.
2. $\angle 1 = \angle 2$; $\angle 3 = \angle 4$.	2. Why?
3. $\triangle ABC \sim \triangle DEC$.	3. Postulate 5.1.

STATEMENTS	REASONS
4. $\frac{AC}{DC}=\frac{BC}{EC}$.	4. Definition 5.4.
5. $\frac{AC-DC}{DC}=\frac{BC-EC}{EC}$.	5. Why?
6. $\frac{AD}{DC}=\frac{BE}{EC}$.	6. Substitution.

Theorem 5.10

The bisector of one angle of a triangle divides the opposite side in the same ratio as the other two sides.

Given: CD bis $\angle C$.

Prove: $\frac{AD}{DB}=\frac{AC}{CB}$

Construction. Through point A construct a line parallel to line DC. Extend side BC to intersect this line at point E.

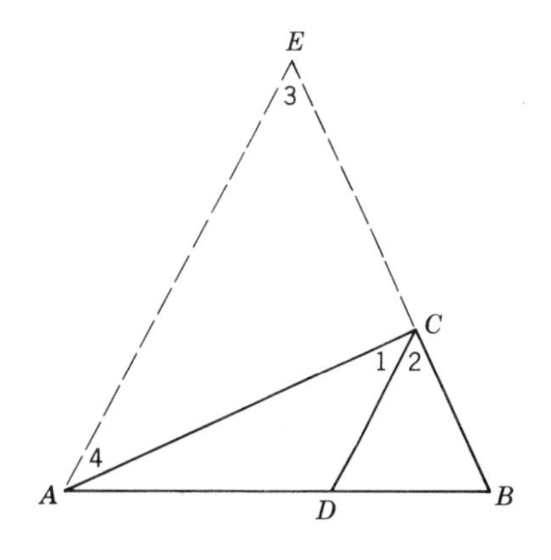

Proof:

STATEMENTS	REASONS
1. CD bis $\angle C$.	1. Given.
2. $\angle 1=\angle 2$.	2. Why?
3. $AE \parallel DC$.	3. By construction.
4. $\angle 3=\angle 2$.	4. Why?
5. $\angle 1=\angle 4$.	5. Why?
6. $\angle 3=\angle 4$.	6. Why?
7. $EC=AC$.	7. Why?
8. $\frac{AD}{DB}=\frac{EC}{CB}$.	8. Theorem 5.9.
9. $\frac{AD}{DB}=\frac{AC}{CB}$.	9. Substitution. (statement 7).

Example 1

Triangles ABC and $A'B'C'$ have dimensions as indicated in the figure. $\angle A = \angle A'$, $\angle B = \angle B'$. Find the measures of sides $A'B'$ and $C'B'$.

Solution: By Postulate 5.1, the two triangles are similar. Therefore corresponding sides are proportional:

$$\frac{AC}{A'C'} = \frac{CB}{C'B'} = \frac{AB}{A'B'}$$

Substituting

$$\frac{6}{9} = \frac{3}{C'B'} = \frac{7}{A'B'}$$

We first solve the proportion

$$\frac{6}{9} = \frac{3}{C'B'}$$

and find that $C'B' = 9/2$.
We then solve the proportion

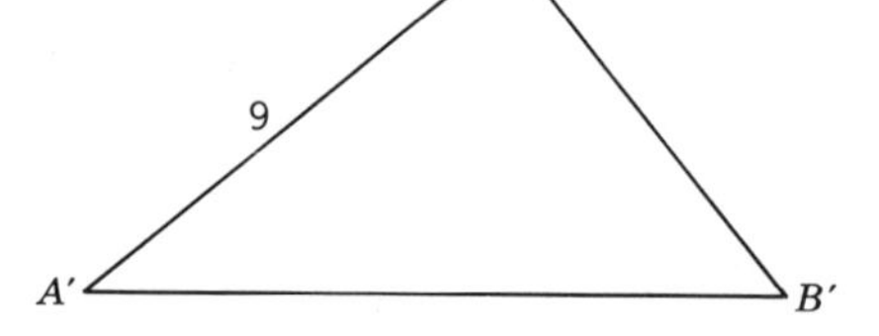

$$\frac{6}{9} = \frac{7}{A'B'}$$

and find that $A'B' = 21/2$.

Example 2

A right triangle has legs of length 6 and 8 inches. CD bisects the right angle. Find the measures of AD and DB.

Solution: By Theorem 5.10, we know that $6/8 = AD/DB$. We write this proportion by addition:

$$\frac{6+8}{8} = \frac{AD+DB}{DB}$$

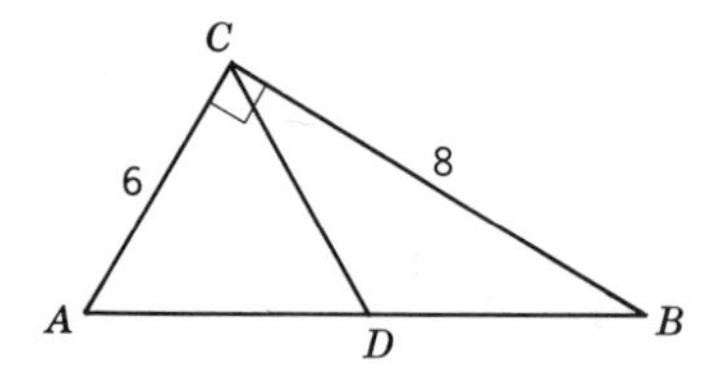

However, $AD + DB = AB$, the hypotenuse of a right triangle. By the Pythagorean theorem, we find that $AB = 10$. (Can you show this?). Our proportion now becomes

$$\frac{14}{8} = \frac{10}{DB}$$

By Theorem 5.1

$$14 \cdot DB = 8 \cdot 10$$

$$DB = \frac{80}{14}$$

$$DB = \frac{40}{7}$$

To find AD, we notice that

$$AD = AB - DB$$

and solve to find that $AD = 30/7$.

EXERCISES

1. *Prove:* the midpoint of the hypotenuse of a right triangle is equidistant from the vertices.
2. *Prove:* a line that bisects one side of a triangle and is parallel to one side of the triangle, must bisect the third side.
3. $AB \parallel DE$, $AD = 4$, $DC = 9$ and $BE = 7$. Find CE.
4. $AB \parallel DE$, $AD = 5$, $DC = 12$ and $CB = 20$. Find CE and EB.

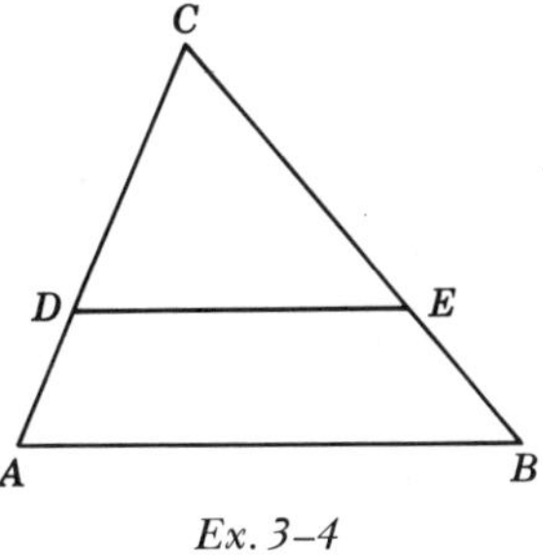

Ex. 3–4

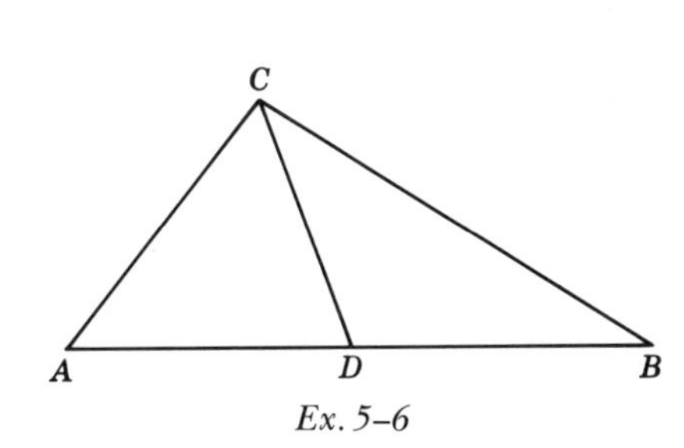

Ex. 5–6

5. CD bisects $\angle C$. If $AC = 4$ and $AD = 2$, find DB.
6. CD bisects $\angle C$. If $AB = 20$ $AC = 10$ and $BC = 5$, find AD and DB.
7. *Prove:* the ratio of the perimeters of two similar triangles is equal to the ratio of a pair of corresponding sides.

8. *Given:* AD and BE intersect at C
$AB \parallel DE$.

Prove: $\frac{AB}{DE} = \frac{AC}{DC}$.

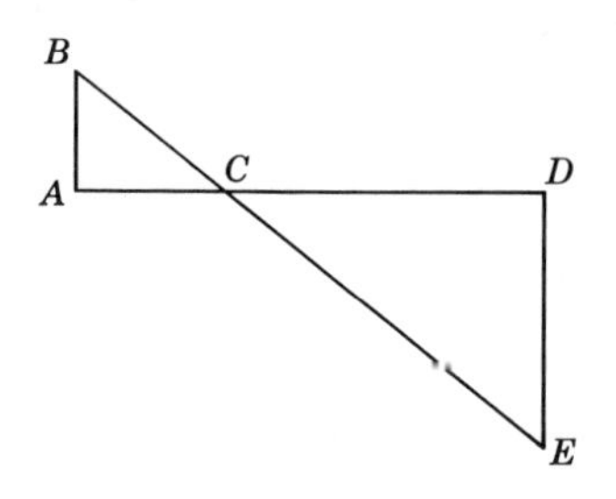

9. Using the information given in problem 8, prove $DE \cdot AC = AB \cdot DC$.

10. *Given:* $\triangle ABE$.
$\angle 1 = \angle A$.

Prove: $\frac{AE}{AC} = \frac{BE}{BD}$.

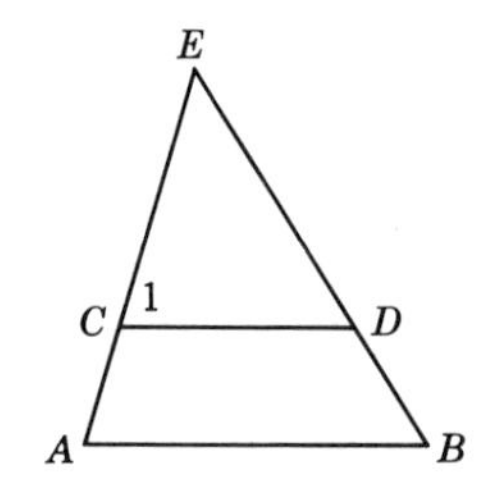

11. Using the information given in problem 10, prove $AC \cdot BE = AE \cdot BD$.

12. *Given:* $\triangle ABC$ and $\triangle ADE$ are right triangles.

Prove: $AD \cdot BC = AB \cdot DE$.

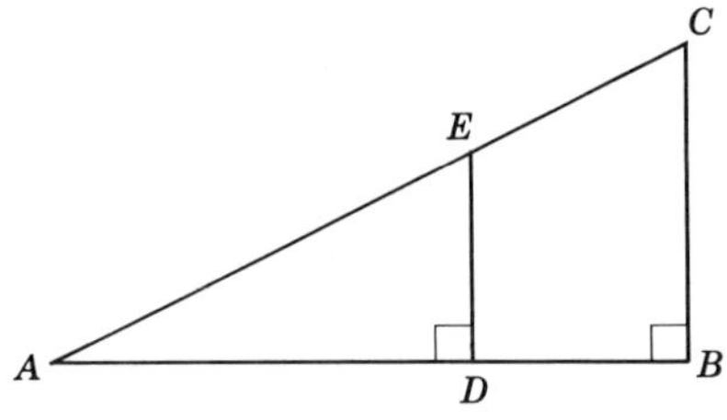

13. Find the measure of the sides EF, AB, and GF.

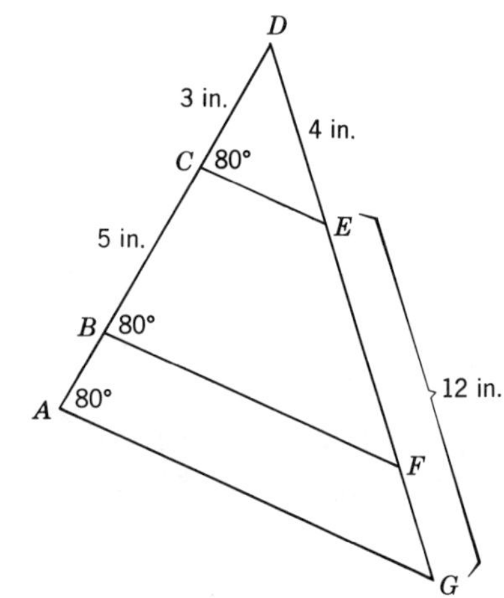

Theorem 5.11

If a line divides two sides of a triangle proportionally, it is parallel to the third side.

Given: $\frac{AD}{DC} = \frac{BE}{EC}$

Prove: $AB \parallel DE$

Construction. Construct line AF parallel to line DE, forming triangle AFC.

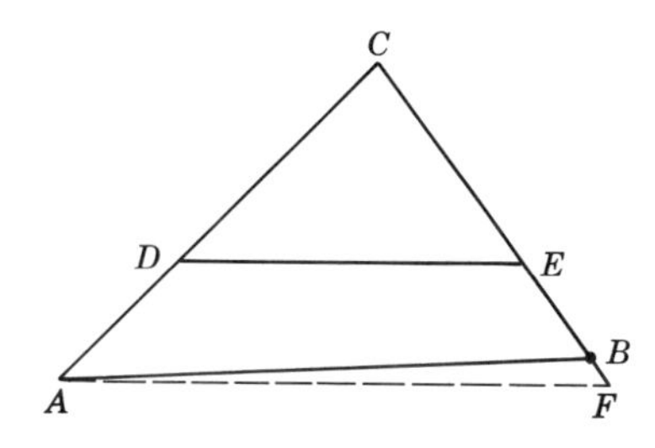

Proof:

STATEMENTS	REASONS
1. $AF \parallel DE$.	1. By construction.
2. $\frac{AD}{DC} = \frac{FE}{EC}$.	2. Theorem 5.9.
3. $\frac{AD}{DC} = \frac{BE}{EC}$.	3. Given.
4. $BE = FE$.	4. Theorem 5.7.
5. F and B must coincide.	5. They are equally distant from E and in the same direction.
6. $AB \parallel DE$.	6. $AF \parallel DE$, and AF and AB are the same line.

Theorem 5.12

All congruent triangles are similar.

The proof of this theorem is left as an exercise.

Theorem 5.13

Two triangles similar to the same triangle are similar to each other.

The proof of this theorem is left as an exercise.

Theorem 5.14

If two triangles have one angle of one equal to one angle of the other and the respective sides including these angles are in proportion, the triangles are similar.

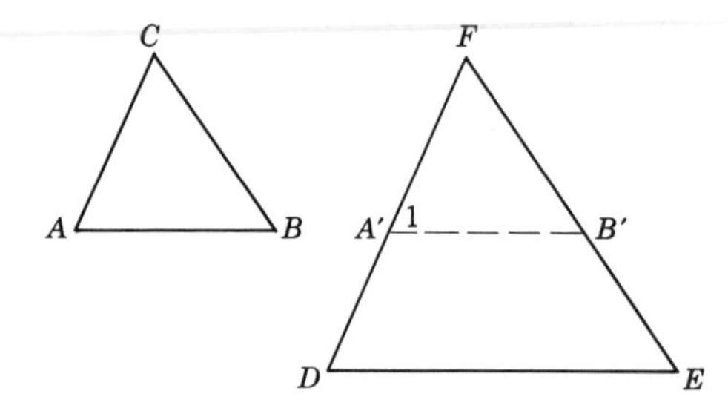

Given: $\angle C = \angle F$.

$$\frac{DF}{AC} = \frac{EF}{BC}.$$

Prove: $\triangle ABC \sim \triangle DEF$.

Construction. Construct point A' on FD so that $FA' = CA$; similarly construct B' on FE so that $FB' = CB$. Draw segment $A'B'$.

Proof:

STATEMENTS	REASONS
1. $\angle C = \angle F$.	1. Given.
2. $A'F = AC$, $B'F = BC$.	2. By construction.
3. $\triangle ABC \cong \triangle A'B'F$.	3. SAS = SAS.
4. $\triangle ABC \sim \triangle A'B'F$.	4. Congruent $\triangle$s are similar.
5. $\frac{DF}{AC} = \frac{EF}{BC}$.	5. Given.
6. $\frac{DF}{A'F} = \frac{EF}{B'F}$.	6. Substitution.
7. $\frac{DA'}{A'F} = \frac{EB'}{B'F}$.	7. A proportion may be written by subtraction.
8. $A'B' \parallel DE$.	8. Theorem 5.11.
9. $\angle D = \angle 1$.	9. Why?
10. $\angle F = \angle F$.	10. Reflexive law.
11. $\triangle A'B'F \sim \triangle DEF$.	11. Why?
12. $\triangle ABC \sim \triangle DEF$.	12. Two triangles similar to the same triangle are similar to each other.

Theorem 5.15

If three sides of one triangle are in proportion to the three corresponding sides of a second triangle, the triangles are similar.

Given: $\frac{DF}{AC} = \frac{EF}{BC} = \frac{DE}{AB}$

Prove: $\triangle DEF \sim \triangle ABC$.

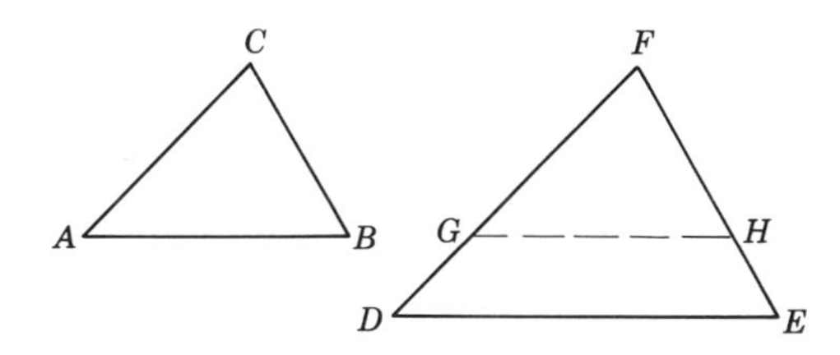

Construction. On sides DF and EF mark off points G and H so that $AC = GF$ and $BC = HF$.

Proof:

STATEMENTS	REASONS
1. $\angle F = \angle F$.	1. Why?
2. $\frac{DF}{AC} = \frac{EF}{BC}$.	2. Given.
3. $AC = GF$; $BC = HF$.	3. By construction.
4. $\frac{DF}{GF} = \frac{EF}{HF}$.	4. Substitution.
5. $\triangle DEF \sim \triangle GHF$.	5. Theorem 5.14.
6. $\frac{DF}{GF} = \frac{DE}{GH}$.	6. Why?
7. $\frac{DF}{AC} = \frac{DE}{GH}$.	7. Substitution.
8. $\frac{DF}{AC} = \frac{DE}{AB}$.	8. Given.
9. $AB = GH$.	9. Theorem 5.7 applied to statements 7 and 8.
10. $\triangle ABC \cong \triangle GHF$.	10. SSS = SSS
11. $\triangle ABC \sim \triangle GHF$.	11. Why?
12. $\triangle ABC \sim \triangle DEF$.	12. Theorem 5.13.

Theorem 5.16

If the altitude is drawn on the hypotenuse in a right triangle, the hypotenuse is to either arm as that arm is to the segment of the hypotenuse adjacent to the arm.

Given: Right triangle ABC
$CD \perp AB$

Prove: $\frac{AB}{AC} = \frac{AC}{AD}$

Proof:

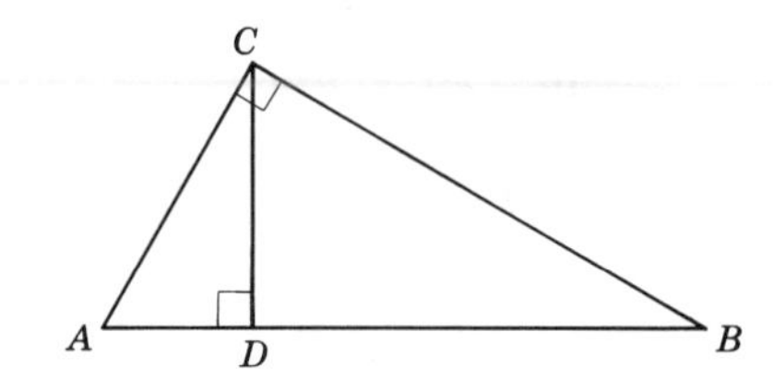

STATEMENTS	REASONS
1. Right $\triangle ABC$; $CD \perp AB$.	1. Given.
2. $\angle A = \angle A$.	2. Reflexive law.
3. $\angle ADC = \angle ACB$.	3. Why?
4. $\triangle ABC \sim \triangle ACD$.	4. Why?
5. $\frac{AB}{AC} = \frac{AC}{AD}$.	5. Why?

By similar reasoning, it is easy to prove that $AB/BC = BC/BD$.

In Chapter 4 we proved the Pythagorean theorem. Many people in the past have enjoyed attempting to find new proofs of this theorem. Many such proofs exist; the following is common to many textbooks. A third proof is given in Appendix I.

Theorem 4.5

In a right triangle, the square of the hypotenuse is equal to the sum of the squares of the other two sides.

Given: Right triangle ABC
$CD \perp AB$

Prove: $a^2 + b^2 = c^2$

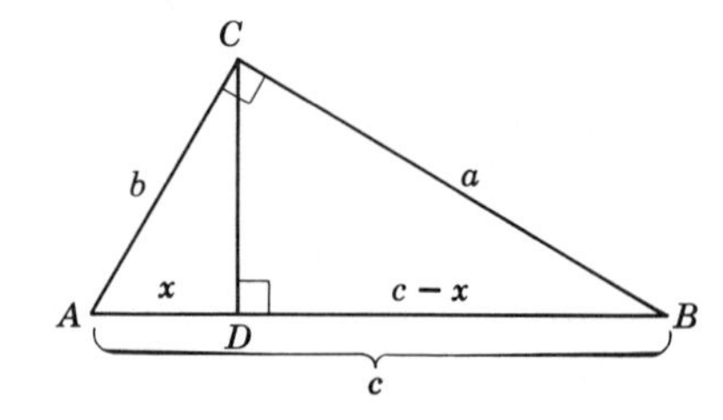

Second Proof:

STATEMENTS	REASONS
1. Rt. $\triangle ABC$ with $CD \perp AB$.	1. Given.
2. $\frac{c}{b} = \frac{b}{x}$; $\frac{c}{a} = \frac{a}{c-x}$.	2. Theorem 5.16.
3. $b^2 = cx$; $a^2 = c^2 - cx$	3. Theorem 5.1.
4. $a^2 + b^2 = c^2$.	4. Equals added to equals are equal.

If the second and third terms of a proportion are equal, the second or third term is called the *mean proportional* between the first and fourth terms.

Theorem 5.17

The altitude on the hypotenuse of a right triangle is the mean proportional between the segments of the hypotenuse.

Given: Right triangle ABC

Prove: $\frac{AD}{CD} = \frac{CD}{BD}$

Construction. Construct CD perpendicular to AB.

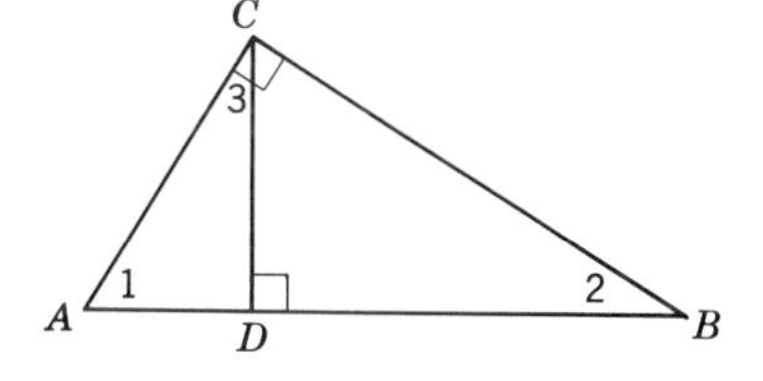

Proof:

STATEMENTS	REASONS
1. Rt. $\triangle ABC$.	1. Given.
2. $CD \perp AB$.	2. By construction.
3. $\angle 1$ comp. $\angle 2$.	3. Why?
4. $\angle 1$ comp. $\angle 3$.	4. Why?
5. $\angle 2 = \angle 3$.	5. Complements of the same angle are equal.
6. $\angle ADC = \angle CDB$.	6. Perpendicular lines meet and form equal adjacent angles.
7. $\triangle ADC \sim \triangle CDB$.	7. Postulate 5.1.
8. $\frac{AD}{CD} = \frac{CD}{BD}$.	8. Why?

The concept of similarity of triangles may be generalized to include all polygons.

Definition 5.5 Two *polygons* are *similar* if, and only if, all pairs of corresponding angles are equal and all pairs of corresponding sides are in proportion.

By definition of similar polygons, we know that the figures at the right are similar if $\angle A = \angle A'$, $\angle B = \angle B'$, $\angle C = \angle C'$, $\angle D = \angle D'$, $\angle E = \angle E'$ and that

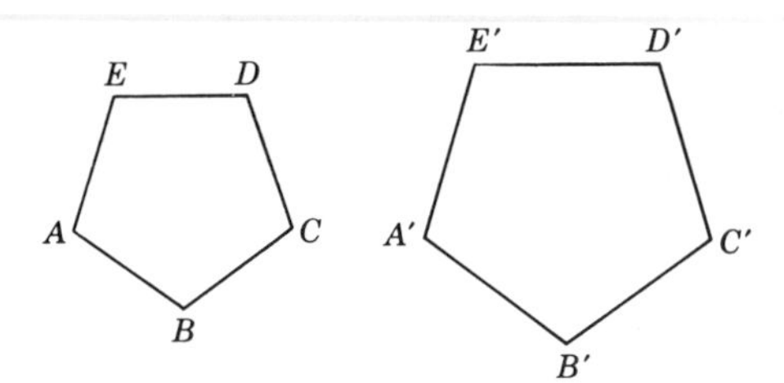

$$\frac{AB}{A'B'} = \frac{BC}{B'C'} = \frac{CD}{C'D'} = \frac{DE}{D'E'} = \frac{EA}{E'A'}$$

EXERCISES

1. $AD = 9$ inches and $DB = 4$ inches. Find sides AC, BC and DC.
2. $AB = 12$ and $CD = 6$. Find AD, DB and AC.
3. $AD = 4$ and $DB = 9$. Find the area of $\triangle ADC$ and $\triangle BDC$.

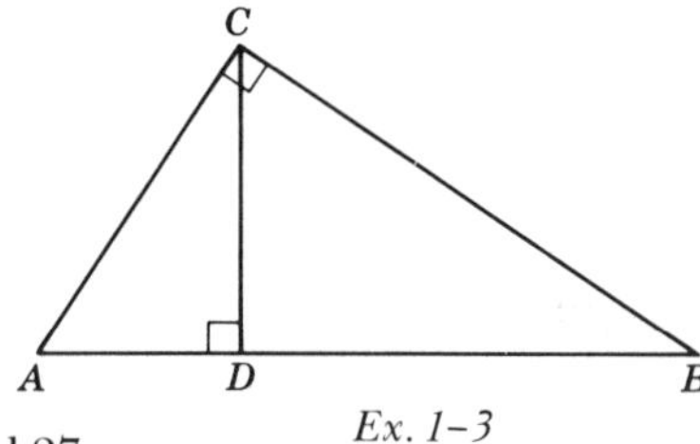

Ex. 1–3

4. Find the mean proportional between 3 and 27.
5. If three successive terms of a proportion are 6, 8, and 32, find the fourth proportional (fourth term).
6. Are triangles with sides of 3, 5, 7 and 12, 20, 28 similar? Why?
7. *Prove:* two isosceles triangles are similar if the vertex angle of one is equal to the vertex angle of the other.
8. *Prove:* the line segments joining the midpoints of the sides of a triangle form a triangle similar to the given triangle.
9. *Prove:* in similar triangles, the ratio of two corresponding medians is equal to the ratio of two corresponding sides.
10. *Prove:* in similar triangles, the ratio of two corresponding angle bisectors is equal to the ratio of two corresponding sides.
11. *Prove:* in similar triangles, the ratio of two corresponding altitudes is equal to the ratio of two corresponding sides.
12. *Prove:* the perimeters of two similar polygons have the same ratio as a pair of corresponding sides.

13. *Prove:* all squares are similar.
14. Are all rectangles similar? Why?
15. Prove Theorem 5.12.
16. Prove Theorem 5.13.
17. *Prove:* if lines are drawn perpendicular to the sides of a given angle, the lines will meet and form an angle supplementary to the given angle.
18. While walking in the woods in northern Wisconsin you come upon a huge White Pine tree that has been estimated to be over 400 years old. A small pine tree, 3 feet tall, casts a shadow of 2 feet, and the large White Pine casts a shadow of 116 feet. Use your knowledge of similar triangles and compute the height of the large tree.
19. Suppose that we wish to find the distance, d, across a river but are unable to swim to the other side and measure the distance directly. By using similar triangles, we can find the distance easily. Find a marker on the opposite side of the river. A few feet from the river and on your side place a stake in the ground directly across from the marker. Measure a given distance on a line perpendicular to the line determined by the marker and the stake and form similar triangles as in the figure. From your side of the river, you can measure sides x, y, and z. With these measures how would you determine the distance, d?

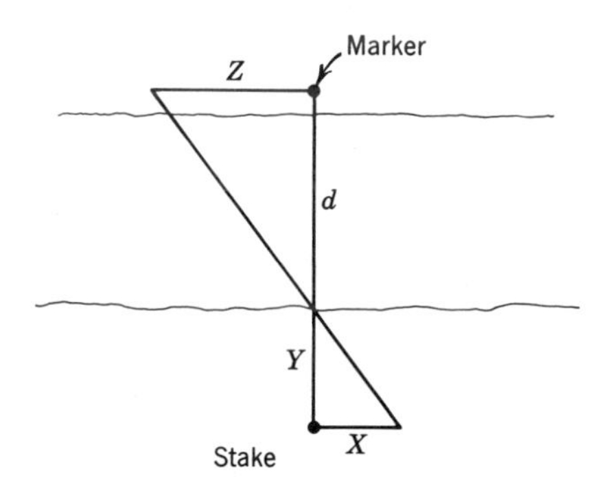

20. Find the distance across the river in problem 19 if $x = 2$ feet, $y = 6$ feet, and $z = 50$ feet.

Review Test

Classify the following statements as true or false.

1. If $a/b = c/d$, then $ad = bc$.
2. If $a/b = c/d$, and $a = c$, then $b = d$.

3. If $a/b = c/d$, then $d/b = c/a$.
4. If $a/b = c/d$, then $(a+c)/(b+d) = a/c$.
5. If two triangles are similar, each pair of corresponding parts must be equal.
6. In $\triangle ABC$, $BD = 20/3$.

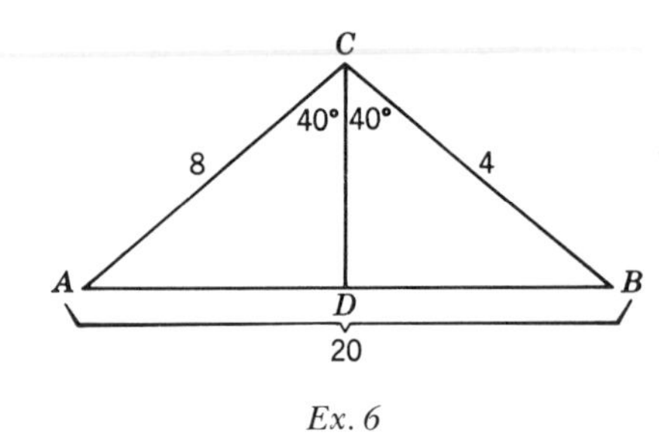

Ex. 6

7. Triangles with sides of 6, 8, 10 and 15, 20, 25 are similar.
8. The mean proportional to 4 and 6 is 5.
9. If an altitude is drawn upon the hypotenuse of a right triangle, the arms of the triangle are in proportion to the segments of the hypotenuse.
10. A triangle with sides of 5, 12, and 13 is a right triangle.
11. If two arms of a right triangle are 15 and 20, the measure of the hypotenuse is 25.
12. All right triangles are similar.
13. If two isosceles triangles have one pair of corresponding base angles equal, the triangles are similar.
14. If two triangles are congruent, they are similar.
15. A proportion is a statement that two ratios are equal.
16. In two similar triangles, the ratio of two corresponding angles equals the ratio of two corresponding sides.
17. In two similar triangles, the ratio of the areas equals the ratio of two corresponding sides.
18. Adding a constant to both numerator and denominator of a fraction does not affect the value of the fraction.
19. If an acute angle of one right triangle is complementary to an acute angle of another right triangle, the triangles are similar.
20. If two angles of a triangle are complementary, the triangle is a right triangle.

6

Circles and More on Similarity

In this chapter, we prove some interesting theorems about circles, their arcs, and related angles. We also review some algebraic techniques in our development of this material.

Definition 6.1 A *circle* is a set of points in the same plane equidistant from a fixed point called its *center*. A line segment drawn from the center of the circle to one of the points on the circle is called a *radius* of the circle.

The plural of "radius" is "radii."

From the definition of a circle it is obvious that all radii of a circle are equal. The center of a circle is usually denoted by the letter O. Anytime a circle has an interior point that is named point O, we will assume point O is the center of the circle.

Definition 6.2 A line segment joining two points on a circle is called a *chord* of the circle. A chord that passes through the center of the circle is called a *diameter* of the circle.

Definition 6.3 A line that has one and only one point of intersection with a circle is called a *tangent* to the circle. Their common point is called a *point of tangency*.

Theorem 6.1

The length of a diameter is twice the length of a radius.

The proof of this theorem is left as an exercise.

Postulate 6.1 A line drawn from the center of a circle to a point of tangency is perpendicular to the tangent passing through the point of tangency. Also, if a line is perpendicular to a radius at its intersection with its circle, the line is tangent to the circle.

The above postulate gives us an easy method of constructing a tangent to a circle at a point on the circle. Suppose we are given a circle and we wish to

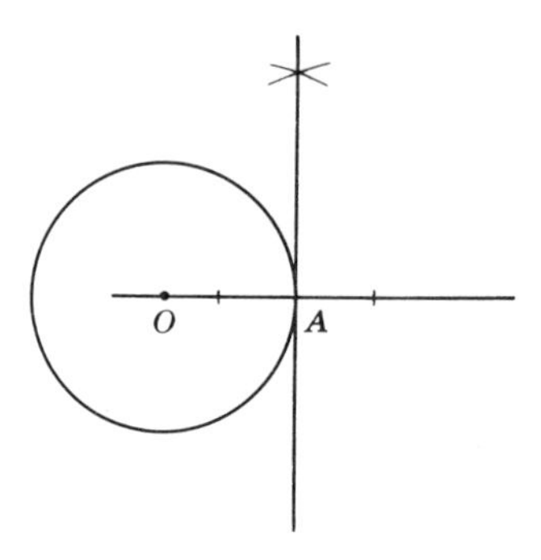

construct a tangent to the circle through point A. We first draw a radius to point A and then construct a perpendicular to the radius at point A. This line will be the desired tangent.

Definition 6.4 A portion of a circle is called an *arc* of the circle. A *semicircle* is an arc of a circle whose endpoints lie on the extremities of a diameter of the circle. An arc greater than a semicircle is called a *major arc*. An arc less than a semicircle is called a *minor arc*.

Arcs are named by using three points on the arc. In Fig. 6.1, arc ABC (usually written $\overset{\frown}{ABC}$) has its endpoints at A and C. Notice that the chord AC divides the circle into two different arcs; one above AC and one below AC.

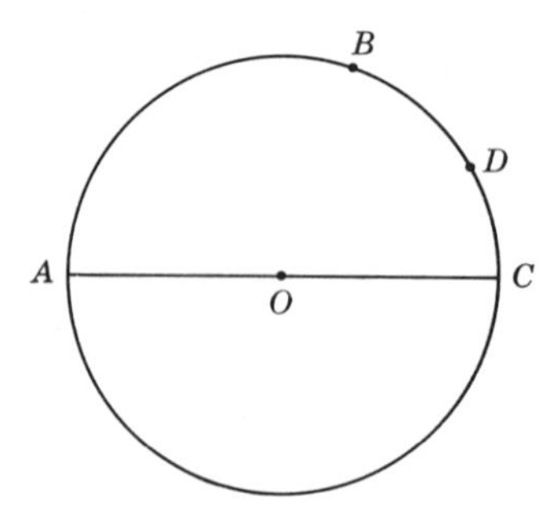

Fig. 6.1

The purpose of the middle letter in naming an arc is to identify which of the two arcs we are considering. In Fig. 6.1, $\widehat{ABC}$ is a semicircle lying above the diameter AC. $\widehat{BDC}$ is a minor arc and $\widehat{BAC}$ is a major arc. If we economize on the notation and use only two letters to name an arc, we will always be considering a minor arc.

Postulate 6.2 Two circles are *congruent* if, and only if, their radii or diameters are equal.

Definition 6.5 An angle whose vertex is at the center of a circle and whose sides are radii is called a *central angle*.

Definition 6.6 The number of degrees in the arc intercepted by a central angle is equal to the number of degrees in the central angle. This number is called the *measure of the arc*.

Notice that the measure of an arc is *not* its length.

As a special case of Definition 6.6, the measure of a semicircle is 180°. It is obvious that equal central angles have equal arcs.

Definition 6.7 An angle whose vertex is on the circle and whose sides are chords of the circle is called an *inscribed angle*.

In the figure to the right $\angle a$ is a central angle. $\angle b$ is an inscribed angle.

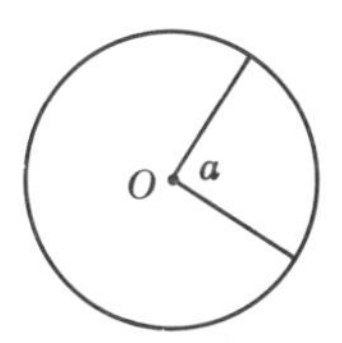

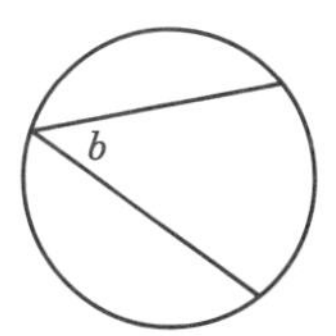

We can now prove an important theorem about the measure of inscribed angles. Theorem 6.2 will be the key to many of the other theorems in this chapter.

Theorem 6.2

The measure of an inscribed angle is equal to one-half the measure of its intercepted arc.

In the proof of this theorem, it is necessary to look at three separate cases. Case I considers the possibility of the center of the circle being on the inscribed angle. Case II considers the possibility of the center of the circle being on the interior of the angle. Case III considers the possibility of the center of the circle being on the exterior of the angle.

CASE I

Given: $\angle ABC$ is an inscribed angle.
O is the center of the circle.

Prove: $\angle B = \frac{1}{2}\widehat{AC}$.

Construction. Draw line segment OC.

Proof:

STATEMENTS	REASONS
1. $\angle ABC$ is an inscribed angle. O is the center of the circle.	1. Given.
2. $\angle 1 = \widehat{AC}$.	2. A central angle has the same measure as its intercepted arc.
3. $OB = OC$.	3. All radii of the same circle are equal.
4. $\angle B = \angle C$.	4. Why?
5. $\angle B + \angle C = \angle 1$.	5. Why?
6. $\angle B + \angle B = 2\angle B = \angle 1$.	6. Substitution.
7. $\angle B = \frac{1}{2}\angle 1$.	7. Why?
8. $\angle B = \frac{1}{2}\widehat{AC}$.	8. Why?

CASE II

Given: $\angle ABC$ is an inscribed angle.
O is the center of the circle.

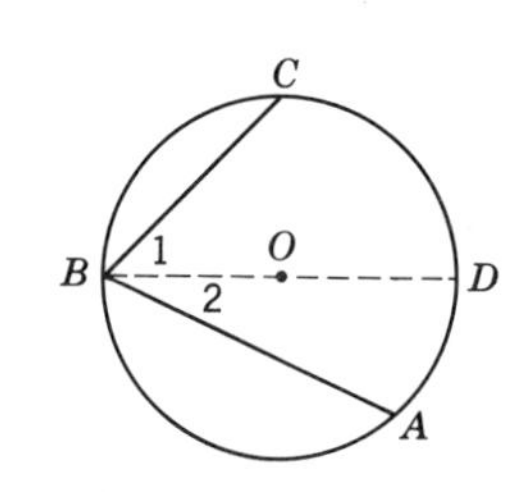

Prove: $\angle ABC = \frac{1}{2}\widehat{AC}$.

Construction. Draw line segment OB intersecting the circle at point D.

Proof:

STATEMENTS	REASONS
1. $\angle ABC$ is an inscribed angle. O is the center of the circle.	1. Given.
2. $\angle 1 = \frac{1}{2}\widehat{DC}$; $\angle 2 = \frac{1}{2}\widehat{AD}$.	2. Why?

STATEMENTS	REASONS
3 $\angle 2 + \angle 1 = \angle ABC$.	3. Why?
4. $\widehat{AD} + \widehat{DC} = \widehat{AC}$.	4. Why?
5. $\angle 2 + \angle 1 = \frac{1}{2}\widehat{AD} + \frac{1}{2}\widehat{DC}$.	5. Why?
6. $\angle 2 + \angle 1 - \frac{1}{2}(\widehat{AD} + \widehat{DC})$	6. Distributive law.
7. $\angle ABC = \frac{1}{2}\widehat{AC}$.	7. Substitution.

CASE III

Given: $\angle ABC$ is an inscribed angle.
O is the center of the circle.

Prove: $\angle B = \frac{1}{2}\widehat{AC}$.

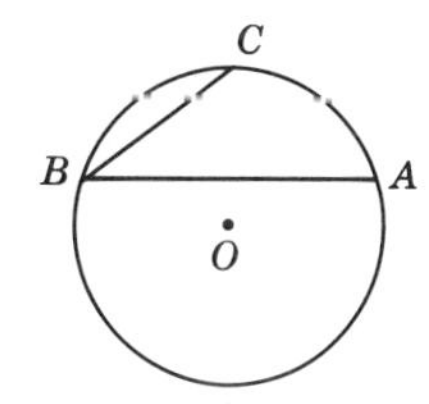

The proof of Case III is left as an exercise.

Theorem 6.3

An angle formed by a tangent and a chord is equal to one-half the measure of the intercepted arc.

Given: $\angle ABC$ is formed by a tangent and a chord.

Prove: $\angle ABC = \frac{1}{2}\widehat{AB}$.

Construction. Draw diameter DB.

Proof:

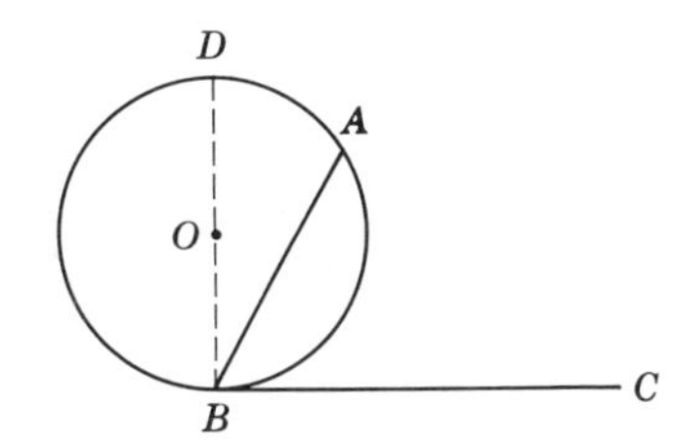

STATEMENTS	REASONS
1. $\angle ABC$ formed by a tangent and chord.	1. Given.
2. DB is a diameter.	2. By construction.
3. $\angle ABC = \angle DBC - \angle DBA$.	3. Why?
4. $\angle DBC = 90°$.	4. Why?
5. $\angle DBA = \frac{1}{2}\widehat{DA}$.	5. Why?
6. $\angle ABC = 90° - \frac{1}{2}\widehat{DA}$.	6. Why?

STATEMENTS	REASONS
7. $2\angle ABC = 180° - \overset{\frown}{DA}$.	7. Why?
8. $\overset{\frown}{DAB} = 180°$.	8. Why?
9. $2\angle ABC = \overset{\frown}{DAB} - \overset{\frown}{DA}$.	9. Why?
10. $2\angle ABC = \overset{\frown}{AB}$.	10. Substitution.
11. $\angle ABC = \frac{1}{2}\overset{\frown}{AB}$.	11. Why?

Theorem 6.4

If two chords intersect within a circle, each angle formed is equal to one-half the sum of its intercepted arc and the intercepted arc of its vertical angle.

Given: AB and DC are chords intersecting at point E.

Prove: $\angle 1 = \frac{1}{2}(\overset{\frown}{BC} + \overset{\frown}{AD})$.

Construction. Draw segment AC.

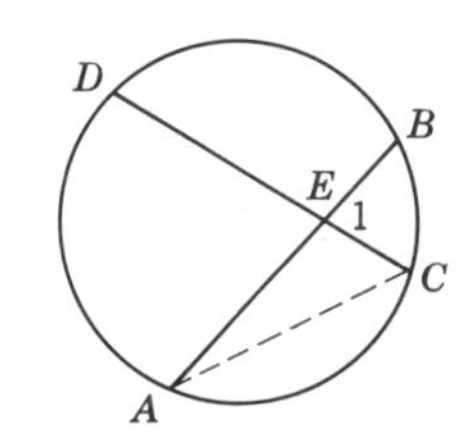

Proof:

STATEMENTS	REASONS
1. AB and DC are chords intersecting at E.	1. Given.
2. $\angle A = \frac{1}{2}\overset{\frown}{BC}$; $\angle C = \frac{1}{2}\overset{\frown}{AD}$.	2. Why?
3. $\angle 1 = \angle A + \angle C$.	3. Why?
4. $\angle 1 = \frac{1}{2}\overset{\frown}{BC} + \frac{1}{2}\overset{\frown}{AD}$.	4. Substitution.
5. $\angle 1 = \frac{1}{2}(\overset{\frown}{BC} + \overset{\frown}{AD})$.	5. Distributive law.

Definition 6.8 A line that intersects a circle in two points is called a *secant*.

Theorem 6.5

An angle formed by the intersection of two secants outside a circle is equal to one-half the difference of its intercepted arcs.

Given: Secants CE and CA

Prove: $\angle C = \frac{1}{2}(\overset{\frown}{AE} - \overset{\frown}{BD})$.

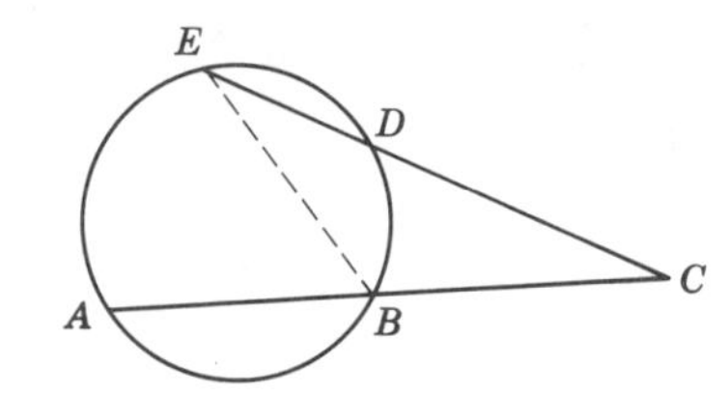

The proof of Theorem 6.5 is left as an exercise.

Theorem 6.6

An angle formed by the intersection of a tangent and a secant outside a circle is equal to one-half the difference of the intercepted arcs.

Given: Tangent CD and secant AC.

Prove: $\angle C = \frac{1}{2}(\widehat{AD} - \widehat{BD})$

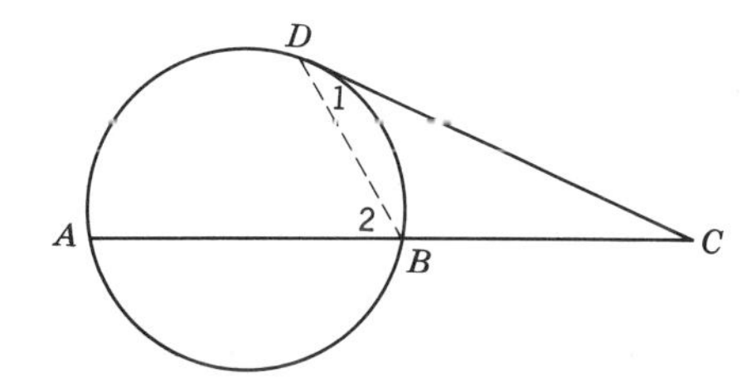

The proof of Theorem 6.6 is left as an exercise.

Example 1

Given: $\widehat{AE} = 80°$
$\angle C = 20°$

Find: The measures of $\angle 1$, $\angle 2$ and $\widehat{BD}$.

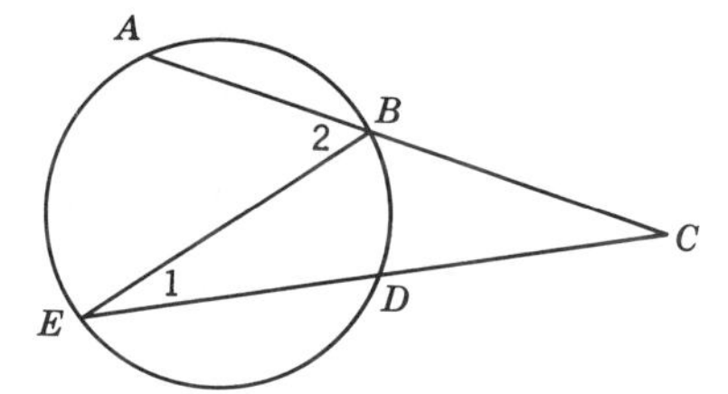

Solution: Since $\angle 2$ is an inscribed angle which intercepts an arc of 80°, $\angle 2$ must be one-half of that arc. Therefore $\angle 2$ equals 40°. By Theorem 6.5, we know that $\angle C = \frac{1}{2}(\widehat{AE} - \widehat{BD})$. Substituting the given values,

$$20° = \frac{1}{2}(80° - \widehat{BD})$$
$$40° = 80° - \widehat{BD}$$
$$\widehat{BD} = 40°$$

Since $\angle 1$ is an inscribed angle which intercepts this 40° arc, $\angle 1$ must be one-half of 40°, or 20°.

EXERCISES

1. Prove Theorem 6.1.
2. Prove case III of Theorem 6.2.
3. Prove Theorem 6.5.
4. Prove Theorem 6.6.
5. Find the measure of $\angle A$, $\angle O$.

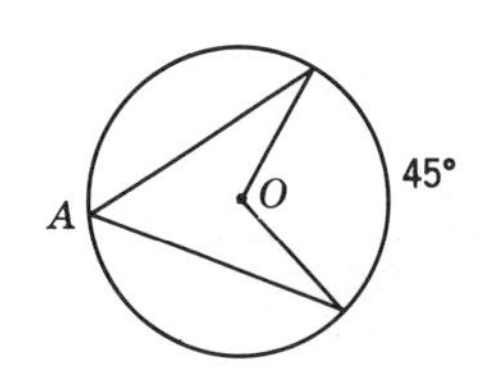

Ex. 5

6. Find the measures of $\angle 1$ and $\angle 2$.

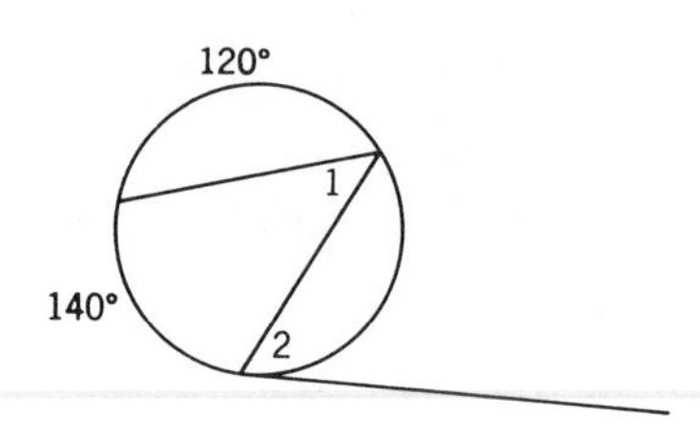

Ex. 6

7. Find the measure of $\angle 1$.

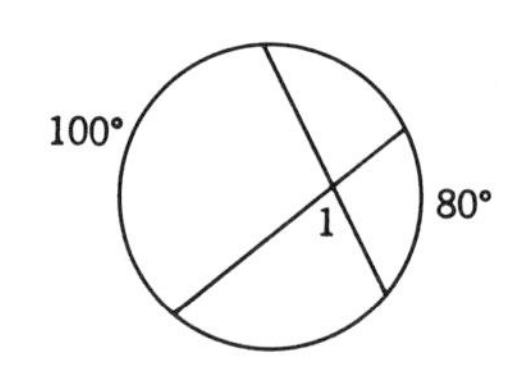

Ex. 7

8. If $\widehat{BD} = 60°$ and $\widehat{AE} = 100°$, find the measure of $\angle C$.

9. If $\angle C = 40°$ and $\widehat{BD} = 40°$, find the measure of $\widehat{AE}$.

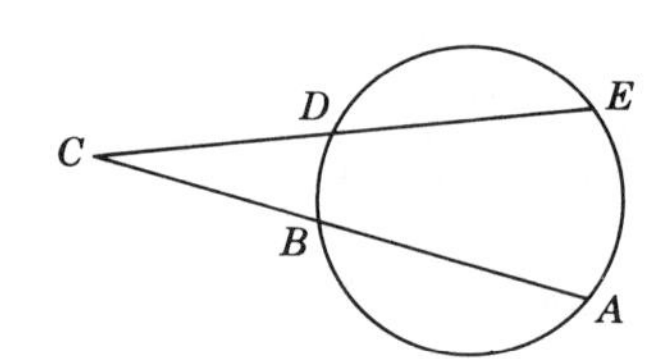

Ex. 8–9

10. AC is a tangent to circle O and AB is a secant. Find the measure of $\widehat{CD}$.

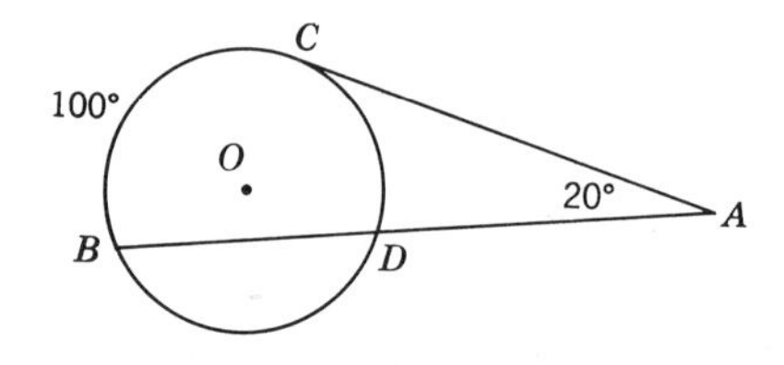

Ex. 10

11. $\angle ABC = 35°$ and B is a point of tangency. Find the measure of $\widehat{AD}$.

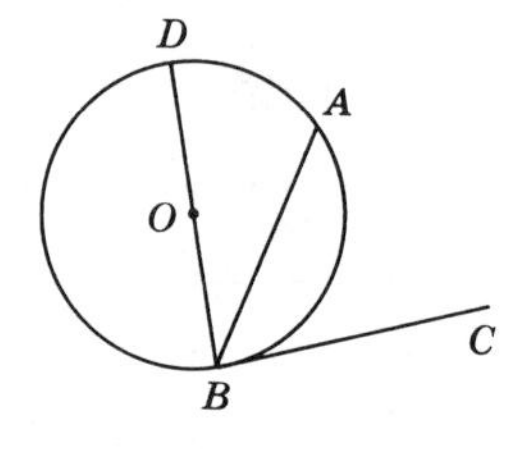

Ex. 11

12. $AB \parallel CD$, C is a point of tangency and $\widehat{AC} = 130°$. Find the measure of $\widehat{BC}$.

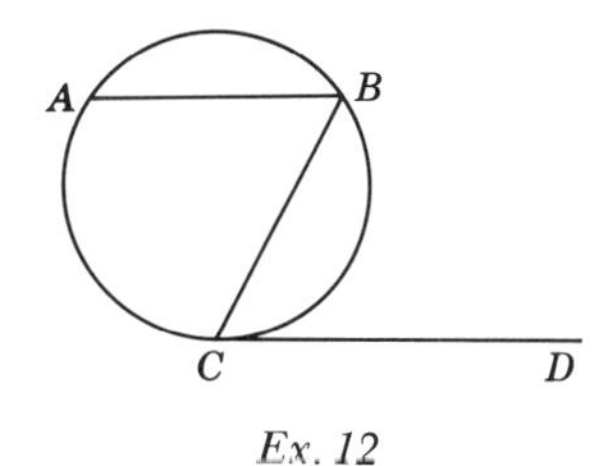

Ex. 12

13. $\angle 1 = 40°$, $\angle 2 = 10°$. Find the measures of $\widehat{AD}$ and $\widehat{BC}$.

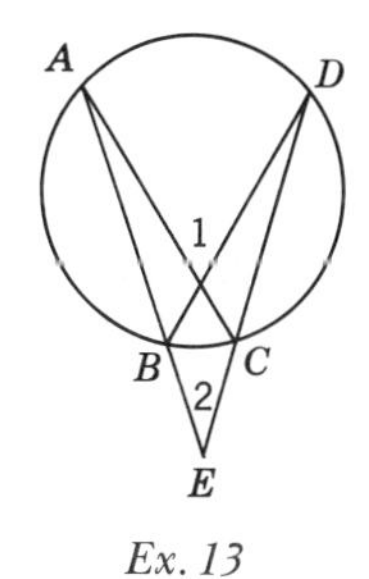

Ex. 13

14. $\angle 1 = 55°$, $\angle D = 40°$. Find the measures of $\widehat{CD}$ and $\angle B$.

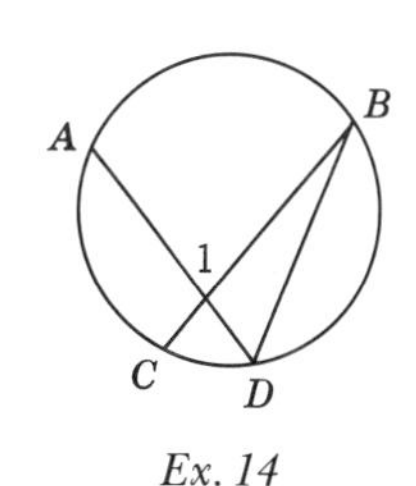

Ex. 14

15. AB and BC are tangents to a circle. $\angle 1 = 70°$. Find the measure of $\angle B$.

16. Explain why equal central angles have equal arcs.

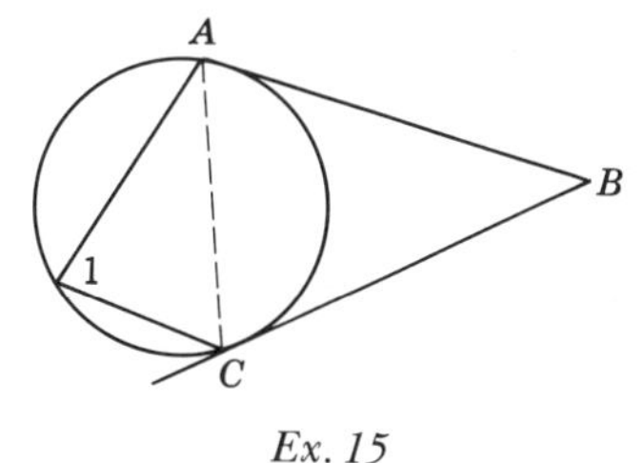

Ex. 15

We are now prepared to prove several additional theorems about circles and their chords, tangents and secants.

Theorem 6.7

A line drawn from the center of a circle perpendicular to a chord bisects the chord and its arc.

Given: O is the center of the circle.
OD is perpendicular to AB.

Prove: OD bisects chord AB.
OD bisects arc AB.

Construction. Draw radii OA and OB.

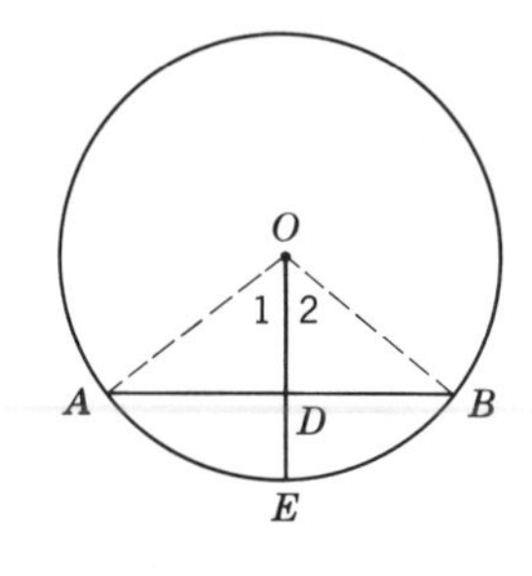

Proof:

STATEMENTS	REASONS
1. O is the center of the circle; $OD \perp AB$.	1. Given.
2. OA and OB are radii.	2. By construction.
3. $OA = OB$.	3. Why?
4. $OD = OD$.	4. Why?
5. $\triangle OAD \cong \triangle OBD$.	5. hs = hs.
6. $AD = DB$; $\angle 1 = \angle 2$.	6. cpcte.
7. $\widehat{AE} = \widehat{EB}$.	7. Equal central angles have equal arcs.
8. OD bisects $\widehat{AB}$.	8. Why?
9. OD bisects AB.	9. Why?

Theorem 6.8

A line drawn from the center of a circle to the midpoint of a chord or the midpoint of its arc is perpendicular to the chord.

Given: M is the midpoint of chord AB.

Prove: $OM \perp AB$.

Construction. Draw radii OA and OB.

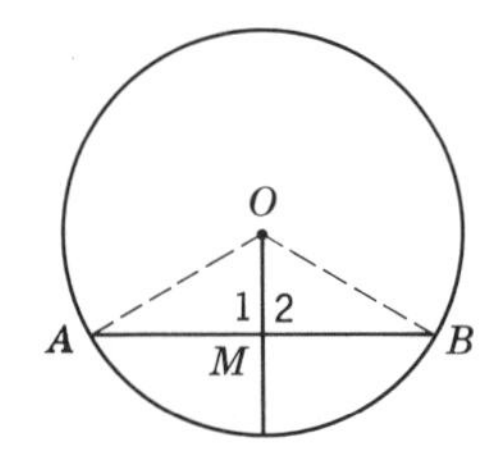

Proof:

STATEMENTS	REASONS
1. M is midpoint of AB.	1. Given.
2. $AM = MB$.	2. Why?
3. $OB = OA$.	3. Why?

STATEMENTS	REASONS
4. $OM = OM$.	4. Why?
5. $\triangle AOM \cong \triangle BOM$.	5. SSS = SSS.
6. $\angle 1 = \angle 2$.	6. cpcte.
7. $OM \perp AB$.	7. Why?

The case where OM bisects $\widehat{AB}$ is left as an exercise.

Theorem 6.9

The perpendicular bisector of a chord passes through the center of the circle.

Given: DE is the perpendicular bisector of AB.

Prove: DE passes through the center of the circle.

Construction. Draw radii OA and OB.

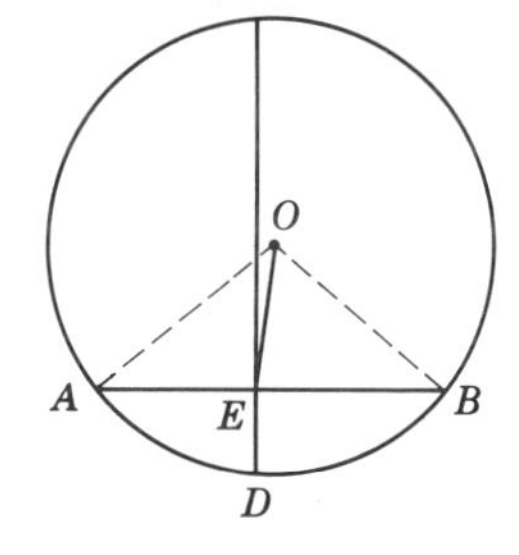

Proof:

STATEMENTS	REASONS
1. $OA = OB$.	1. Why?
2. $\triangle AOB$ is isosceles.	2. Why?
3. DE is the perpendicular bisector of AB.	3. Given.
4. AB is the base of the isosceles $\triangle AOB$.	4. Why?
5. DE passes through the center of the circle.	5. Theorem 2.3.

The above theorem gives us a method of locating the center of an arc or circle. In Fig. 6.2 we first draw a chord, AB, of the arc. Then we construct the perpendicular bisector of AB and call it NM. By the above theorem, NM passes through the center of the circle. Now draw another chord CD and construct its perpendicular bisector RS. RS will pass through the center of the circle, and the intersection of RS and MN will be the center of the circle.

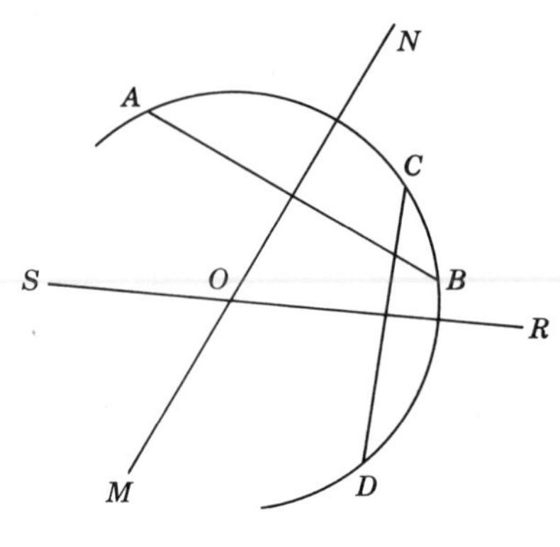

Fig. 6.2

Theorem 6.10

In the same circle or congruent circles equal chords have equal arcs.

Given: $AB = CD$.

Prove: $\widehat{AB} = \widehat{CD}$.

Construction. Draw radii OA, OB, OC, and OD.

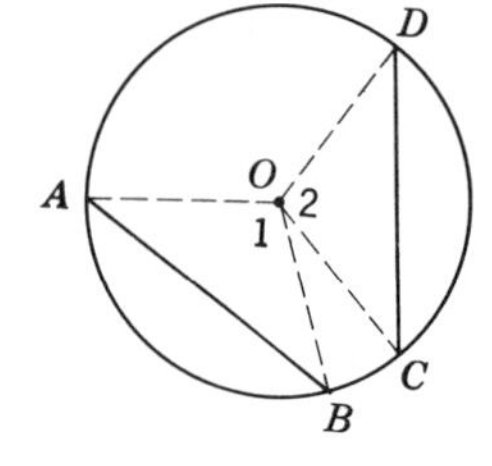

Proof:

STATEMENTS	REASONS
1. $AB = CD$.	1. Given.
2. $OB = OC$; $OA = OD$.	2. Why?
3. $\triangle AOB \cong \triangle DOC$.	3. SSS = SSS.
4. $\angle 1 = \angle 2$.	4. cpcte.
5. $\widehat{AB} = \widehat{DC}$.	5. Why?

The case for congruent circles is left as an exercise.

Theorem 6.11

In the same circle or congruent circles equal arcs have equal chords.

The proof of this theorem is left as an exercise.

Theorem 6.12

In the same circle or congruent circles, equal chords are equidistant from the center.

Given: $AB = CD$.

Prove: AB and CD are equidistant from O.

Construction. Construct $OE \perp AB$ and $OF \perp DC$. OE and OF represent the distances the chords are from the center of the circle.

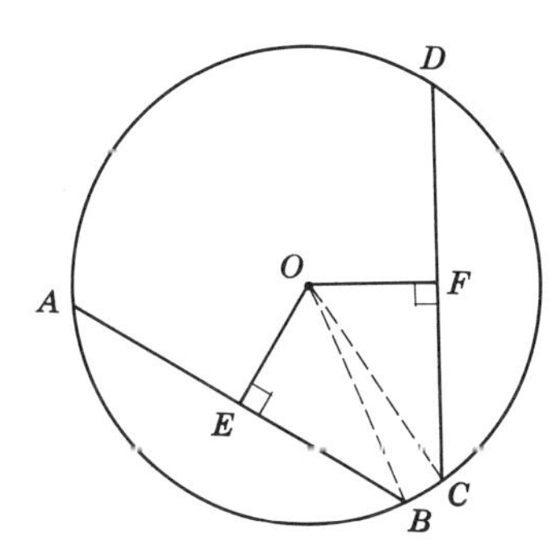

Proof:

STATEMENTS	REASONS
1. $AB = CD$.	1. Given.
2. $OE \perp AB$; $OF \perp DC$.	2. By construction.
3. $AE = EB$; $DF = FC$.	3. Why?
4. $EB = FC$.	4. Halves of equal quantities are equal.
5. $OB = OC$.	5. Why?
6. $\triangle OEB \cong \triangle OFC$.	6. hs = hs.
7. $OE = OF$.	7. cpcte.
8. AB and CD are equidistant from O.	8. Why?

The case for congruent circles is left as an exercise.

Theorem 6.13

In the same circle or congruent circles, chords equidistant from the center are equal.

The proof of Theorem 6.13 is left as an exercise.

EXERCISES

1. Complete the proof of Theorem 6.10.
2. Prove Theorem 6.11.
3. Complete the proof of Theorem 6.12.

4. Prove Theorem 6.13.
5. If OC is perpendicular to AB and $AE = 3$ inches, what is the length of chord AB?
6. If the radius of circle O is 5 inches and chord AB is 3 inches from the center of the circle, what is the measure of the chord AB?
7. If the radius of circle O is 13 inches and $AE = 5$ inches, what is the measure of segment EC?
8. In circle O, $AB = CD$. If $\widehat{BD} = 60°$ and $\widehat{CB} = 20°$, what is the measure of $\widehat{AED}$?
9. Circles O and O' are congruent and $AC = BC$. If $\angle A = 30°$ and $\widehat{BDC} = 210°$, what is the measure of $\widehat{AEC}$?
10. Draw an arc of convenient size. Using only a compass and a straightedge, locate the center of the arc.
11. Draw a rectangle with sides of approximately 2 inches and 3 inches on your paper. Construct a square that has the same area as the rectangle.

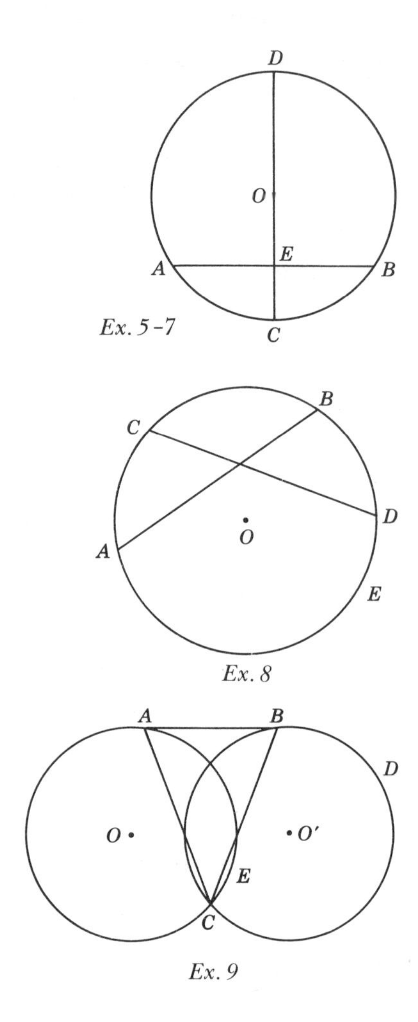

Ex. 5-7

Ex. 8

Ex. 9

The use of the preceding theorems gives us an easy way to construct a tangent to a circle from a point outside the circle, as shown in Fig. 6.3. To construct a tangent to a given circle O from a given point A that is outside the circle, we must find a point C on the circle such that radius OC is perpendicular to the desired tangent CA. To find this point join point O and point A, and bisect this segment at point P. Construct a circle using P as center and PA as radius. The points of intersection of the two circles, points C and D, are the required points. Lines CA and DA are tangent to the original circle. The proof of this construction is left as an exercise.

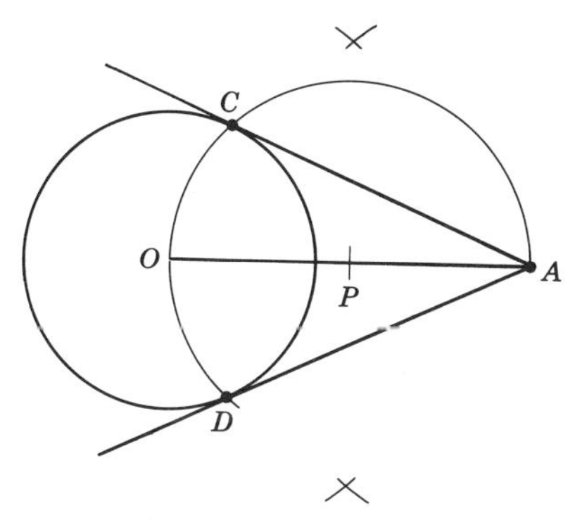

Fig. 6.3

Theorem 6.14

If two tangents are drawn to a circle from a point outside the circle, the tangents are of equal measure.

Given: AB and AC are tangents to circle O.

Prove: $AB = AC$

Construction. Draw radii OB, OC, and segment OA.

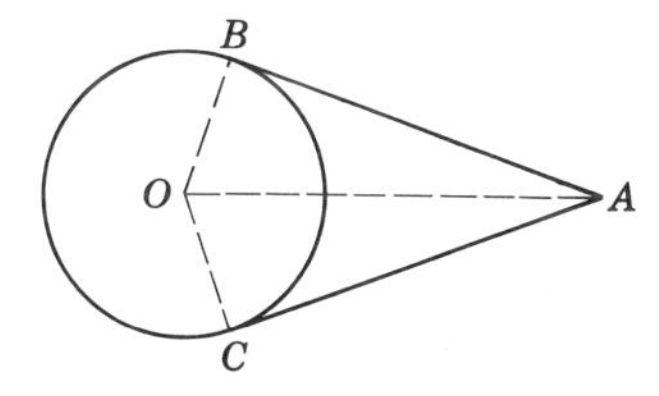

Proof:

STATEMENTS	REASONS
1. AB and AC are tangents to circle O.	1. Given.
2. $OB \perp AB$; $OC \perp CA$.	2. Why?
3. $OC = OB$.	3. All radii of the same circle are equal.
4. $OA = OA$.	4. Reflexive law.
5. $\triangle OAB \cong \triangle OAC$.	5. hs = hs
6. $AB = AC$.	6. cpcte.

Two circles may intersect in two, one, or no points. If two circles intersect in a single point, they are said to be tangent to each other. If one circle is in the interior of the other, except for the point of tangency, the circles are said to be *tangent internally* (see Fig. 6.4*a*). Otherwise they are said to be *tangent externally* (see Fig. 6.4*b*). Two circles that do not intersect may have a common tangent.

If the two circles lie on different sides of their common tangent, the tangent is called an *internal tangent* (see Fig. 6.4*c*). If the circles lie on the same side of their common tangent, the tangent is called an *external tangent* (see Fig. 6.4*d*).

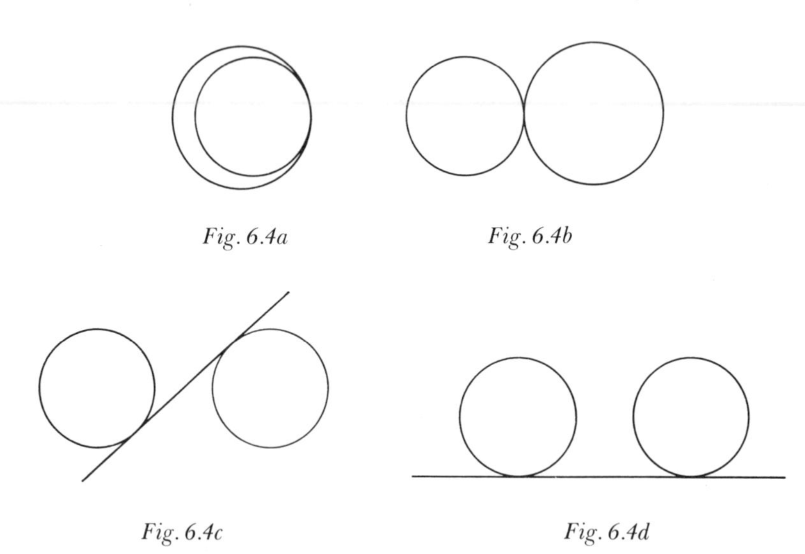

Fig. 6.4a *Fig. 6.4b*

Fig. 6.4c *Fig. 6.4d*

Definition 6.9 The line passing through the centers of two circles is called the *line of centers*.

Definition 6.10 If three or more points lie on the same line, they are said to be *collinear*.

Theorem 6.15

If two circles are tangent either internally or externally, the point of tangency and the centers of the circles are collinear.

Given: Circles O and O' tangent externally with point E as point of tangency. EF is the common internal tangent.

Prove: OO' passes through point E.

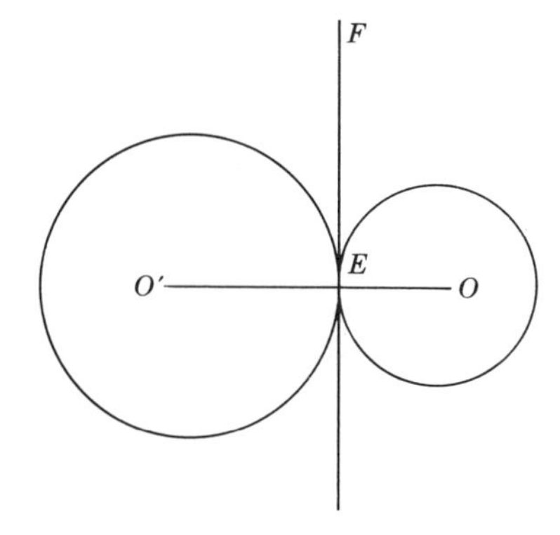

Proof:

STATEMENTS	REASONS
1. Circles O and O' tangent externally with point E as point of tangency. EF is the common internal tangent.	1. Given.
2. Draw $O'E$.	2. Two points determine a line.
3. $O'E \perp EF$ at point E.	3. Why?
4. Draw OE.	4. Same as reason 2.
5. $OE \perp EF$ at point E.	5. Same as reason 3.
6. $O'E$ and OE coincide.	6. There is only one perpendicular to EF at point E.
7. OO' passes through point E.	7. Why?

The case where the circles are tangent internally is left as an exercise.

EXERCISES

1. From a given point outside a circle, how many tangents can be drawn to the circle?
2. From a given point outside a circle, how many secants can be drawn to the circle?
3. How many internal tangents can two circles have? Discuss the possibilities.
4. How many external tangents can two circles have? Discuss the possibilities.
5. Complete the proof of Theorem 6.15.
6. AB and DB are tangents to circle O' and DB and BC are tangents to circle O. *Prove:* $AB = BC$.
7. AB and BC are tangents to circle O. Find the measure of $\angle B$.

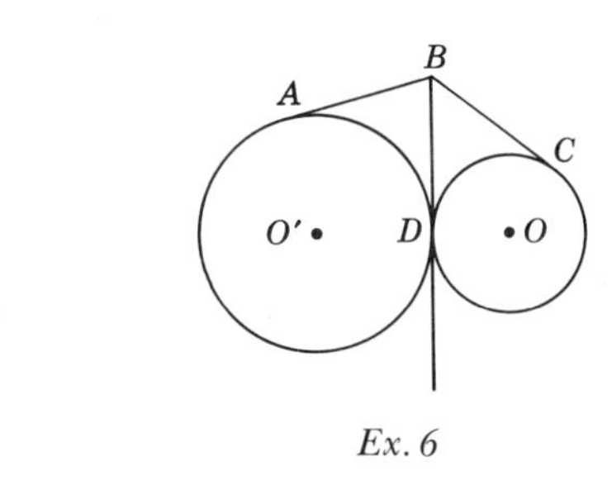

Ex. 6

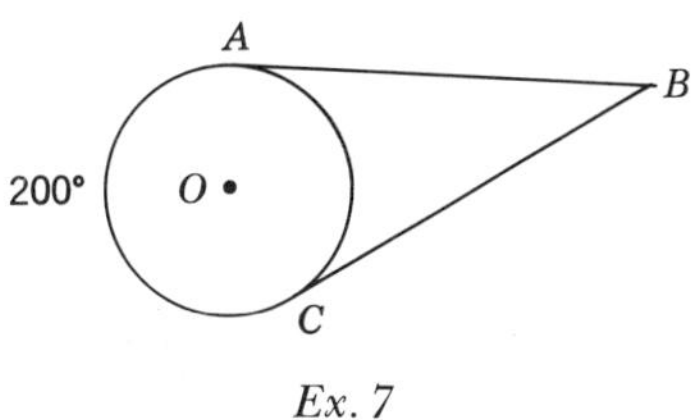

Ex. 7

8. AB and CD are common external tangents to circles O' and O. *Prove:* $AB = CD$.

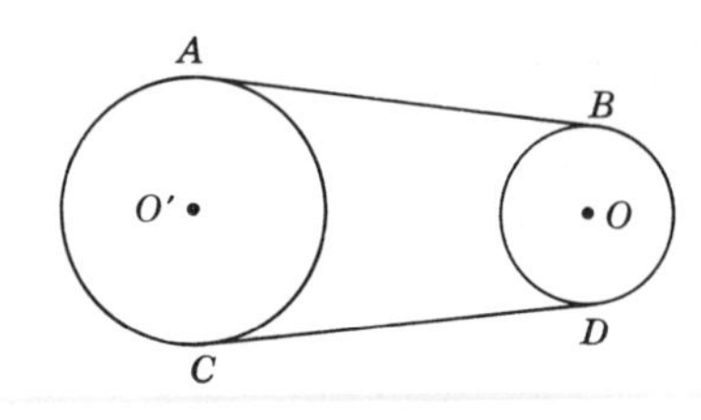

Ex. 8

9. l is tangent to circle O at point E. Find the measure of $\angle 1$; $\angle 2$.

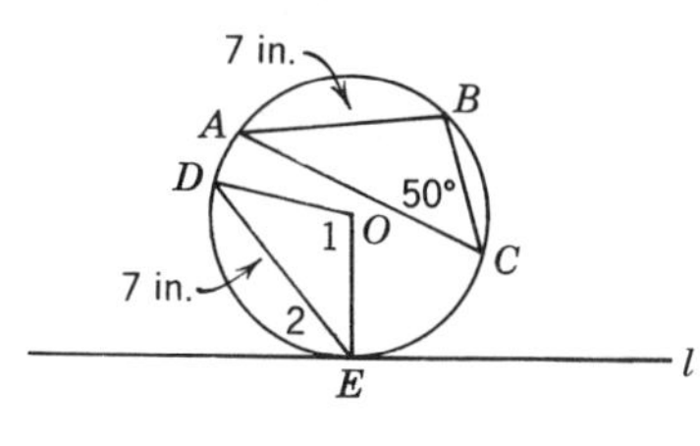

Ex. 9

10. AB is a tangent to circle O and AD is a secant. Find the measure of $\widehat{DC}$.

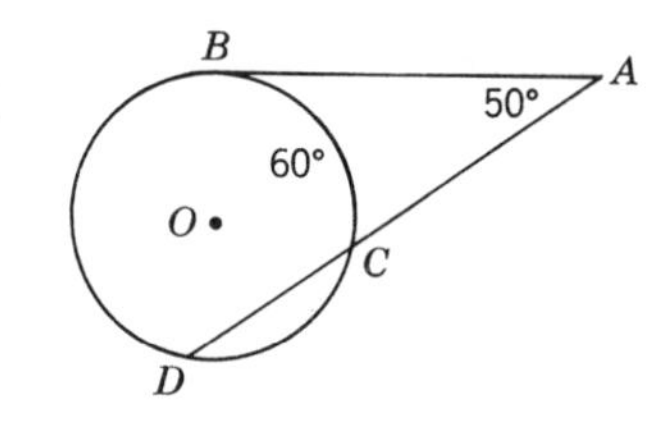

Ex. 10

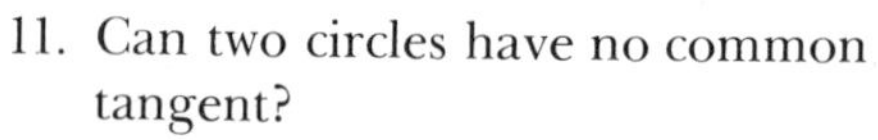

11. Can two circles have no common tangent?
12. Prove that an angle inscribed in a semicircle is a right angle.
13. Prove the construction given on page 128.
14. Complete the proof of Theorem 6.8.

Theorem 6.16

If two circles intersect in two points, their line of centers is the perpendicular bisector of their common chord.

Given: Circle O' intersects circle O at points A and B.

Prove: OO' is perpendicular bisector of AB.

Construction. Draw radii $O'A$, $O'B$, OA, and OB.

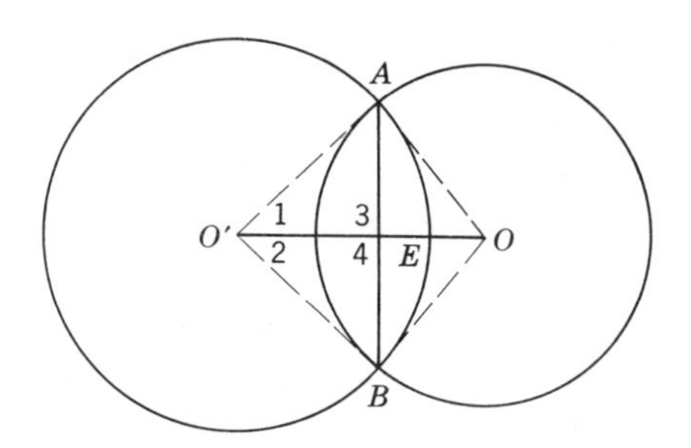

Proof:

STATEMENTS	REASONS
1. Circle O' intersects circle O at points A and B.	1. Given.
2. $O'B = O'A$; $OA = OB$.	2. Why?
3. $OO' = OO'$.	3. Reflexive law.
4. $\triangle O'AO \cong \triangle O'BO$.	4. SSS = SSS.
5. $\angle 1 = \angle 2$.	5. cpcte.
6. $O'E = O'E$.	6. Reflexive law.
7. $\triangle O'AE \cong \triangle O'BE$.	7. SAS = SAS.
8. $BE = AE$.	8. cpcte.
9. $O'O$ bisects AB.	9. Why?
10. $\angle 3 = \angle 4$.	10. cpcte.
11. $O'O$ is perpendicular to AB.	11. Why?
12. OO' is perpendicular bis. of AB.	12. Why?

Theorem 6.17

In a circle, parallel lines intercept equal arcs.

The proof of this theorem is left as an exercise.

Theorem 6.18

If two chords intersect within a circle, the product of the segments of one chord is equal to the product of the segments of the other chord.

Given: Chords AB and CD intersect at E.

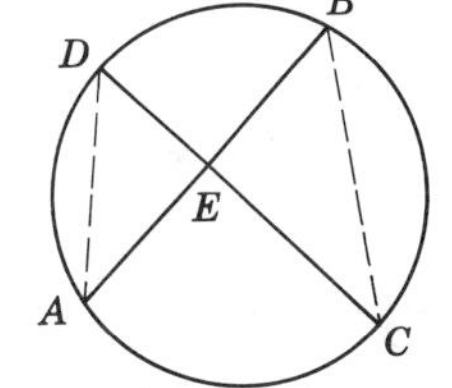

Prove: $(AE)(BE) = (EC)(DE)$

Construction. Draw segments AD and BC.

Proof:

STATEMENTS	REASONS
1. Chords AB and CD intersect at E.	1. Given.
2. $\angle D = \angle B$.	2. Why?
3. $\angle A = \angle C$.	3. Same as reason 2.

STATEMENTS	REASONS
4. $\triangle ADE \sim \triangle CBE$.	4. Why?
5. $\frac{AE}{EC} = \frac{DE}{BE}$.	5. Why?
6. $(AE)(BE) = (EC)(DE)$.	6. In a proportion, the product of the means equals the product of the extremes.

Theorem 6.19

If a secant and a tangent are drawn to a circle, the measure of the tangent is the mean proportional between the secant and its external segment.

Given: AB is tangent to circle O and AD is a secant.

Prove: $\frac{AD}{AB} = \frac{AB}{AC}$.

Construction. Draw segments DB and BC.

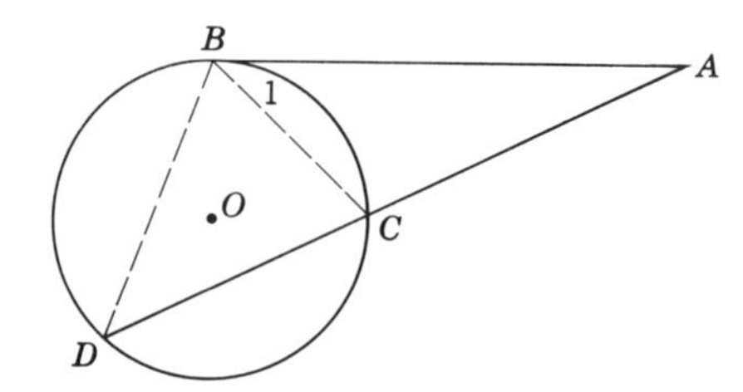

Proof:

STATEMENTS	REASONS
1. AB is a tangent and AD is a secant.	1. Given.
2. $\angle A = \angle A$.	2. Reflexive law.
3. $\angle D = \angle 1$.	3. Why?
4. $\triangle ABD \sim \triangle ACB$.	4. Why?
5. $\frac{AD}{AB} = \frac{AB}{AC}$.	5. Corresponding parts of similar triangles are in proportion.

Theorem 6.20

If two secants are drawn to a circle from a point outside the circle, the products of the secants and their external segments are equal.

Given: AC and AE are secants from point A.

Prove: $(AC)(AB) = (AE)(AD)$.

Construction. Draw segments EB and CD.

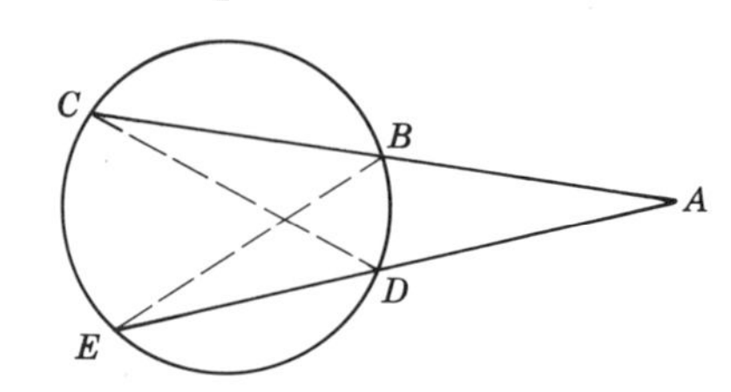

Proof:

STATEMENTS	REASONS
1. AC and AE are secants from point A.	1. Given.
2. $\angle A = \angle A$.	2. Reflexive law.
3. $\angle C = \angle E$.	3. Why?
4. $\triangle ACD \sim \triangle AEB$.	4. Why?
5. $\dfrac{AC}{AE} = \dfrac{AD}{AB}$.	5. Why?
6. $(AC)(AB) = (AE)(AD)$.	6. The product of the means equals the product of the extremes.

If a polygon has its vertices on a circle, the polygon is said to be *inscribed* within the circle. If all sides of the polygon are tangent to a circle, the polygon is said to be *circumscribed* about the circle. In Fig. 6.5*a*, triangle ABC is inscribed within a circle and the circle is said to be circumscribed about the triangle. In Fig. 6.5*b*, triangle DEF is circumscribed about a circle and the circle is said to be inscribed within the triangle.

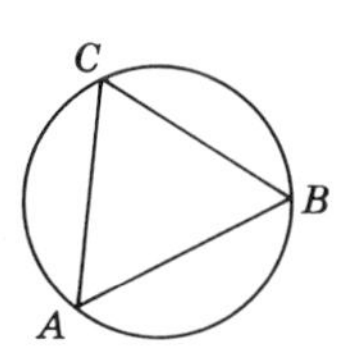

Fig. 6.5a

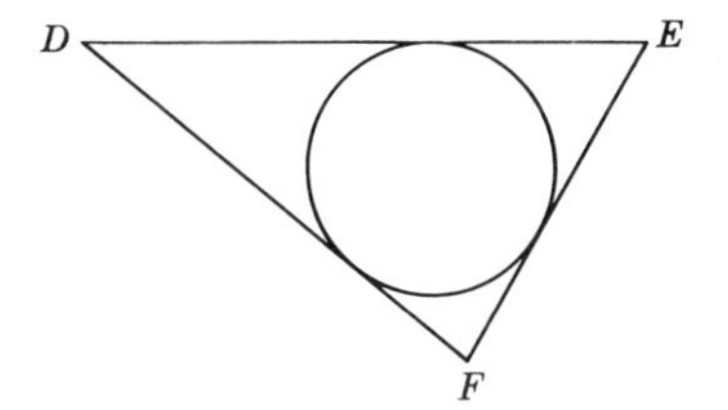

Fig. 6.5b

Theorem 6.21

The opposite angles of an inscribed quadrilateral are supplementary.

Given: $ABCD$ is an inscribed quadrilateral.

Prove: $\angle A$ is supplementary to $\angle C$.

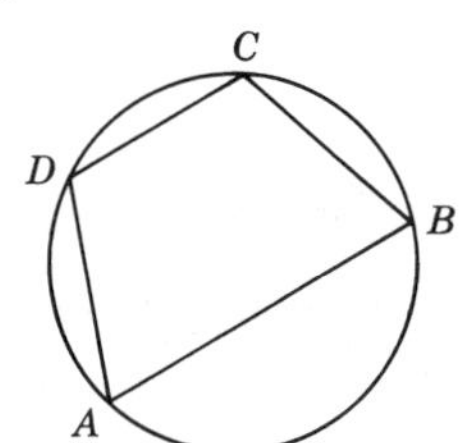

Proof:

STATEMENTS	REASONS
1. *ABCD* is an inscribed quadrilateral.	1. Given.
2. $\angle A = \frac{1}{2}\widehat{DCB}$; $\angle C = \frac{1}{2}\widehat{BAD}$.	2. Why?
3. $\widehat{DCB} + \widehat{BAD} = 360°$.	3. Why?
4. $\angle A + \angle C = \frac{1}{2}\widehat{DCB} + \frac{1}{2}\widehat{BAD}$.	4. Equal quantities added to equal quantities are equal.
5. $\angle A + \angle C = \frac{1}{2}(\widehat{DCB} + \widehat{BAD})$.	5. Distributive law.
6. $\angle A + \angle C = \frac{1}{2}(360°) = 180°$.	6. Substitution.
7. $\angle A$ is supplementary to $\angle C$.	7. Why?

By a similar argument we can show that $\angle D$ is supplementary to $\angle B$.

Theorem 6.22

If a parallelogram is inscribed within a circle, it is a rectangle.

Given: *ABCD* is a parallelogram.

Prove: *ABCD* is a rectangle.

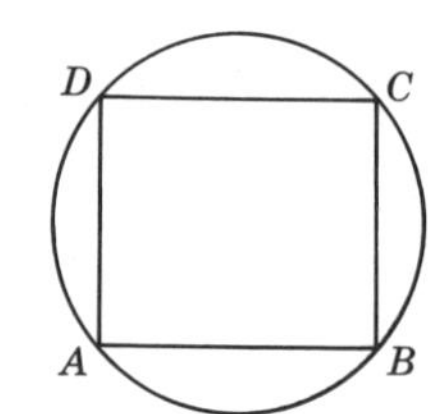

Proof:

STATEMENTS	REASONS
1. *ABCD* is a parallelogram.	1. Given.
2. $\angle A = \angle C$.	2. Why?
3. $\angle A$ is supplementary to $\angle C$.	3. Why?
4. $\angle A = \angle C = 90°$.	4. If two angles are equal and supplementary, each is equal to 90°.
5. *ABCD* is a rectangle.	5. Why?

EXERCISES

1. Prove Theorem 6.17.
2. Find the measure of *EB*.

3. *Prove:* if two tangents are drawn to a circle from a point outside the circle, the line segment joining the center of the circle and the point of intersection of the tangents bisects the angle formed by the tangents.
4. AB is tangent to circle O. If $AD = 12$ and $AC = 4$, find AB.
5. AB is tangent to circle O. If $AD = 20$ and $DC = 12$, find AB.
6. If $AB = 10$ and $AD = 80$, find DC.
7. If $AC = 12$, $BA = 4$, and $AE = 16$, find AD.
8. If $AC = 20$, $BC = 4$, and $AD = 16$, find ED.
9. Find the measure of $\angle C$.
10. Circle O is congruent to circle O'. Prove that $AB = CD$ if AB and CD are common external tangents.
11. A rather surprising result first discovered in 1821 is illustrated by this construction:
    (1) Draw any triangle in the center of a sheet of paper. Make the sides about 2 or 3 inches long.
    (2) Find the midpoints of the three sides and label them A, B, and C.
    (3) Construct the three altitudes of the triangle. Use letters D, E, and F to label the feet of the altitudes.
    (4) If you have been careful, the three altitudes meet in a single point. Bisect each of the three line segments joining the point of intersection of the altitudes to a vertex. Label these midpoints G, H, and I.

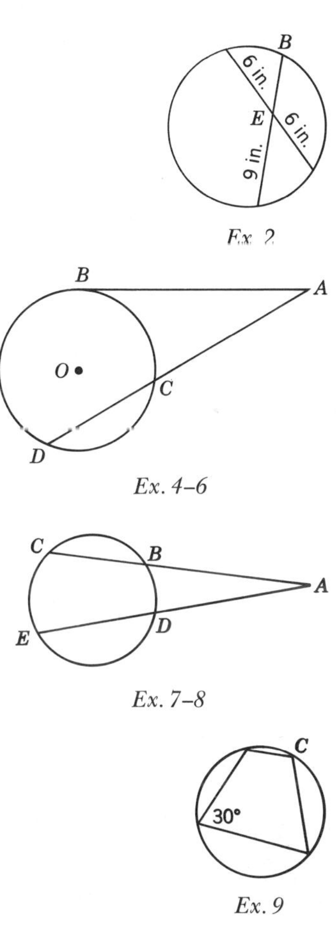

Ex. 2

Ex. 4–6

Ex. 7–8

Ex. 9

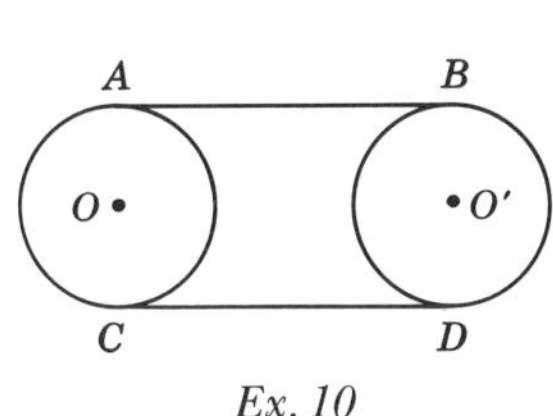

Ex. 10

If your drawing is cluttered, go back and mark more heavily the nine labeled points. What do you notice about these nine points? If you have done the construction accurately, all nine points will lie on a circle, called the *nine-point circle*.

Review Test

Classify the following statements as true or false.

1. A diameter is a chord of a circle.
2. A minor arc is an arc that is less than 90°.
3. An inscribed angle is equal to one-half the measure of its intercepted arc.
4. An inscribed angle is twice as large as a central angle.
5. If two secants are drawn to a circle from a point outside the circle, the products of the segments of one secant equals the product of the segments of the other.
6. Equal chords are equidistant from the center of their circle.
7. If a chord measures 7 inches, its intercepted arc equals 7 inches.
8. Secants drawn to a circle from a point outside the circle are equal.
9. If two circles are tangent internally, they have no internal tangents.
10. Each bisector of a chord passes through the center of the circle.
11. Opposite angles of an inscribed quadrilateral are equal.
12. An inscribed angle can never be an obtuse angle.
13. If two chords intersect within a circle, four central angles will be formed.
14. If a regular pentagon is inscribed within a circle, the circle will be divided into five equal arcs.
15. A regular hexagon is inscribed within a circle. The measure of each arc of the circle is 60°.
16. If two circles intersect, their line of centers bisects their common chord.
17. The opposite angles of a quadrilateral inscribed within a circle are supplementary.
18. The measure of an inscribed angle depends on the radius of the circle.
19. A secant intersects a circle in two points.
20. If two circles are tangent externally, their line of centers is parallel to their common external tangents.

7

Inequalities

In algebra we have studied inequalities. Here we review much of this material and investigate some new topics. In this chapter we regard the term "positive number" as undefined.

Definition 7.1 The symbol "<" means "is less than" and the symbol ">" means "is greater than." If a and b are numbers, $a < b$ if, and only if, there is a positive number p such that $b = a + p$. If a and b are numbers, $a > b$ if, and only if, $b < a$.

Postulate 7.1 Exactly one of the three relations is true: either $a < b$, $a = b$, or $a > b$.

Postulate 7.2 The whole is greater than any of its parts.

Postulate 7.3 The sum of any two sides of a triangle is greater than the third side.

We further assume familiarity with the commutative and associative laws of addition and multiplication. These laws state that for any numbers a, b, and c:

1. $a + b = b + a$; commutative law of addition.
2. $a \cdot b = b \cdot a$; commutative law of multiplication.
3. $(a + b) + c = a + (b + c)$; associative law of addition.
4. $(a \cdot b) \cdot c = a \cdot (b \cdot c)$; associative law of multiplication.

Also assumed are the usual properties of addition and multiplication of signed numbers. We are now prepared to prove many theorems about inequalities.

Theorem 7.1

If the same quantity is added to both sides of an inequality, the sums are unequal and in the same order.
(If $a < b$, then $a + c < b + c$.)

Given: $a < b$.

Prove: $a + c < b + c$.

Proof:

STATEMENTS	REASONS
1. $a < b$.	1. Given.
2. $b = a + p$ where p is a positive number.	2. Definition 7.1.
3. $b + c = a + p + c$.	3. Why?
4. $b + c = a + c + p$.	4. Associative and commutative laws of addition.
5. $a + c < b + c$.	5. Definition 7.1.

Theorem 7.2

If equal quantities are added to unequal quantities, the sums are unequal and in the same order.
(If $a < b$ and $c = d$, then $a + c < b + d$.)

Given: $a < b$; $c = d$.

Prove: $a + c < b + d$.

Proof:

STATEMENTS	REASONS
1. $a < b$; $c = d$.	1. Given.
2. $b = a + p$ where p is a positive number.	2. Definition 7.1.
3. $b + d = a + p + c$.	3. Equal quantities added to equal quantities are equal.
4. $b + d = a + c + p$.	4. Associative and commutative laws of addition.
5. $a + c < b + d$.	5. Definition 7.1.

We notice that Theorems 7.1 and 7.2 allow for the subtraction of the same or equal quantities from both sides of an inequality, because the subtraction of a quantity, a, is simply the addition of the quantity, $(-a)$.

Theorem 7.3

If unequal quantities are subtracted from equal quantities, the differences are unequal and in the opposite order.
(If $a < b$ and $c = d$, then $c - a > d - b$.)

Given: $a < b; c = d$.

Prove: $c - a > d - b$.

Proof:

STATEMENTS	REASONS
1. $a < b; c = d$.	1. Given.
2. $b = a + p$ where p is a positive number.	2. Definition 7.1.
3. $c - (a + p) = d - b$.	3. Why?
4. $c - a - p = d - b$.	4. Distributive law.
5. $c - a = d - b + p$ where p is a positive number.	5. Why?
6. $c - a > d - b$.	6. Why?

Theorem 7.4

If both sides of an inequality are multiplied by a positive number, the products are unequal and in the same order.
(If $a < b$ and c is a positive number, then $ac < bc$.)

Given: $a < b; c > 0$.

Prove: $ac < bc$.

Proof:

STATEMENTS	REASONS
1. $a < b; c > 0$.	1. Given.
2. $b = a + p$ where p is a positive number.	2. Why?

STATEMENTS	REASONS
3. $bc = (a+p)c$.	3. Equal quantities multiplied by equal quantities are equal quantities.
4. $bc = ac + pc$.	4. Distributive law.
5. pc is a positive number.	5. A positive number multiplied by a positive number is a positive number.
6. $ac < bc$.	6. Why?

Theorem 7.5

If both sides of an inequality are multiplied by a negative number, the products are unequal in the opposite order.
(If $a < b$ and c is a negative number, then $ac > bc$.)

Given: $a < b$; $c < 0$.

Prove: $ac > bc$.

Proof:

STATEMENTS	REASONS
1. $a < b, c < 0$.	1. Given.
2. $b = a + p$ where p is a positive number.	2. Definition 7.1.
3. $bc = (a+p)c$.	3. Equal quantities multiplied by equal quantities are equal.
4. $bc = ac + pc$.	4. Distributive law.
5. pc is a negative number.	5. A positive number multiplied by a negative number is a negative number.
6. $ac = bc + (-pc)$.	6. Why?
7. $-pc$ is a positive number.	7. The opposite of a negative number is a positive number.
8. $bc < ac$ or $ac > bc$.	8. Definition 7.1.

Theorem 7.6

The relation "<" is transitive.
(If $a < b$ and $b < c$, then $a < c$.)

Given: $a < b; b < c$.

Prove: $a < c$.

Proof:

STATEMENTS	REASONS
1. $a < b; b < c$.	1. Given.
2. $b = a + p; c = b + q$ where p and q are positive numbers.	2. Why?
3. $c = a + p + q$.	3. Substitution.
4. $p + q$ is a positive number.	4. Why?
5. $a < c$.	5. Definition 7.1.

Theorem 7.7

If unequal quantities are added to unequal quantities of the same order, the sums are unequal quantities and in the same order.
(If $a < b$ and $c < d$, then $a + c < b + d$.)

Given: $a < b$ and $c < d$.

Prove: $a + c < b + d$

Proof:

STATEMENTS	REASONS
1. $a < b; c < d$.	1. Given.
2. $b = a + p;\ d = c + q$ where p and q are positive numbers.	2. Definition 7.1.
3. $b + d = a + p + c + q$.	3. Equal quantities added to equal quantities are equal quantities.
4. $b + d = a + c + p + q$.	4. Associative and commutative laws of addition.
5. $p + q$ is a positive number.	5. Why?
6. $a + c < b + d$.	6. Definition 7.1.

Theorem 7.8

In a triangle an exterior angle is greater than either nonadjacent interior angle.

Given: $\triangle ABC$; $\angle C > 0°$.

Prove: $\angle 1 > \angle 2$.

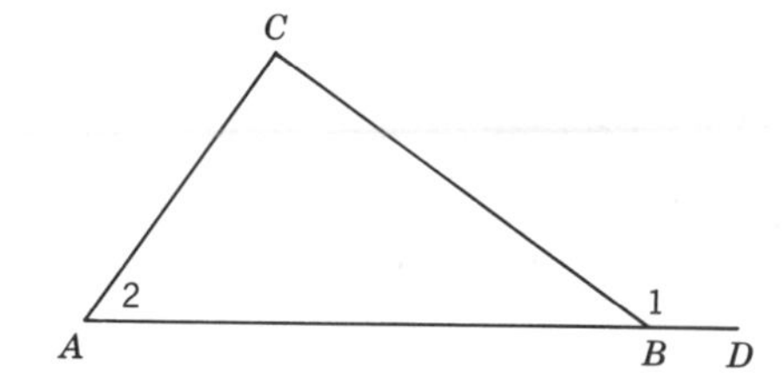

Proof:

STATEMENTS	REASONS
1. $\angle C > 0°$.	1. Given.
2. $\angle 1 = \angle 2 + \angle C$.	2. In a triangle an exterior angle equals the sum of the nonadjacent interior angles.
3. $\angle C = \angle 1 - \angle 2$.	3. Why?
4. $\angle 1 - \angle 2 > 0$.	4. Substitution.
5. $\angle 1 > \angle 2$.	5. Why?

By a similar argument $\angle 1$ can be shown to be greater than $\angle C$.

Example 1

Solve the inequality $x + 7 < 2x + 2$.

$$x + 7 < 2x + 2$$

By Theorem 7.2, we may add "-2" to both sides without affecting the direction of the inequality:

$$x + 7 + (-2) < 2x + 2 + (-2)$$
$$x + 5 < 2x$$

Similarly, we may add "$-x$" to both sides.

$$x + 5 + (-x) < 2x + (-x)$$
$$5 < x$$

Hence, our original inequality will be satisfied if x is any number greater than 5.

Example 2

Solve the inequality $x(x-3) < x^2+6$.

$$x(x-3) < x^2+6$$

By the distributive law,

$$x^2-3x < x^2+6$$

By Theorem 7.2, we may add $-x^2$ to both sides:

$$-3x < 6$$

Theorem 7.5 allows us to multiply by $-\frac{1}{3}$, provided that we reverse the direction of the inequality:

$$x > -2.$$

Example 3

Solve the inequality $x(x+1) > (x^2+x+2)$.

$$x^2+x > x^2+x+2$$

$$x > x+2$$

$$0 > 2$$

What a strange result, $0 > 2$! Let us think a moment about what we have done. Solving this inequality means finding all those values of x that will make the original inequality a true statement. But the original inequality reduces to the statement, $0 > 2$. Since no value of x can possibly make 0 greater than 2, we conclude *no* value of x will solve the original inequality.

EXERCISES

Solve the inequalities in problems 1 to 7. Be able to give a reason for each step.

1. $3x < 2x+4$.
2. $3(x-4) > 6x-2$.
3. $3x/2 < 5+x$.
4. $x(x^2-7) > 3x+x^3-4$.
5. $7x-3 > \frac{2}{3}x+7$.
6. $2x-\frac{1}{3} > x+3$.
7. $2x^2 < x(x-7)+x(x+12)$.

8. *Prove:* an exterior angle associated with a base angle of an isosceles triangle is unequal to the vertex angle.
9. *Prove:* the diagonal of a rectangle is larger than any side of the rectangle.
10. *Prove:* the shortest distance between two points is a straight line.

*11. *Prove:* the relation "$>$" is transitive.

Theorem 7.9

If two sides of a triangle are unequal, the angles opposite these sides are unequal in the same order.

Given: $BC > AC$.

Prove: $\angle BAC > \angle B$.

Construction. On side CB construct a segment CT so that $CT = CA$.

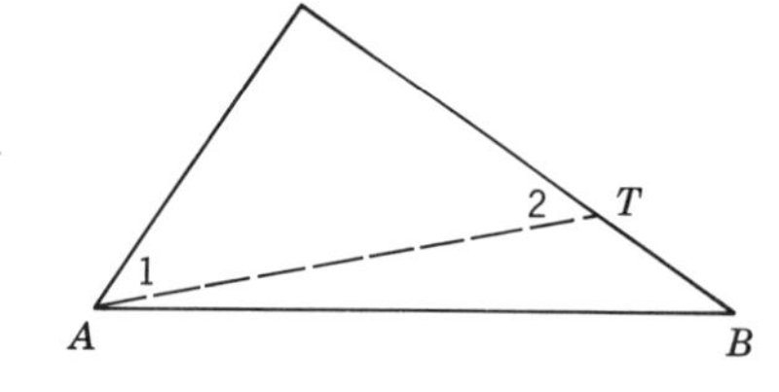

Proof:

STATEMENTS	REASONS
1. $CT = CA$.	1. By construction.
2. $\angle 1 = \angle 2$.	2. Why?
3. $\angle BAC > \angle 1$.	3. Postulate 7.2.
4. $\angle BAC > \angle 2$.	4. Substitution.
5. $\angle 2 > \angle B$.	5. Theorem 7.8.
6. $\angle BAC > \angle B$.	6. Transitive property of the relation "$>$".

Theorem 7.10

If two angles of a triangle are unequal, the sides opposite these angles are unequal in the same order.

Given: $\angle A > \angle B$.

Prove: $BC > AC$.

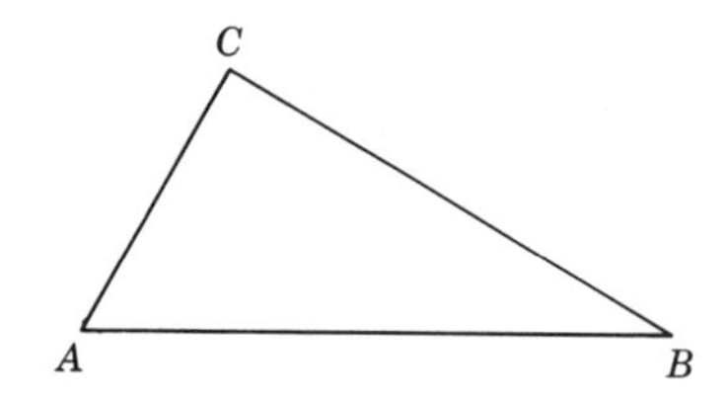

Plan: The proof is indirect. We first assume $BC = AC$, and then assume $BC < AC$ and show that either of these assumptions leads to a contradiction. $BC > AC$ is then the only remaining possibility.

Proof:

STATEMENTS	REASONS
1. $BC = AC$.	1. By assumption.
2. $\angle A = \angle B$.	2. Why?
3. $\angle A > \angle B$.	3. Given.
4. Statement 1 is false.	4. It leads to a contradiction (statements 2 and 3).
5. $BC < AC$.	5. By assumption.
6. $\angle A < \angle B$.	6. Theorem 7.9.
7. Statement 5 is false.	7. It leads to a contradiction (statements 3 and 6).
8. $BC > AC$ or $BC = AC$ or $BC < AC$.	8. Postulate 7.1.
9. $BC > AC$.	9. It is the only remaining possibility.

Theorem 7.11

If two sides of one triangle are equal to two sides of a second triangle and the included angle of the first is greater than the included angle of the second, then the third side of the first triangle is greater than the third side of the second triangle.

Plan: The proof of this theorem will be considered in three cases depending on whether an endpoint of constructed line segment CG falls outside, on, or inside a given triangle.

CASE I (point G falls outside $\triangle ABC$)

Given: $AC = DF$.
$BC = EF$.
$\angle C > \angle F$.

Prove: $AB > DE$.

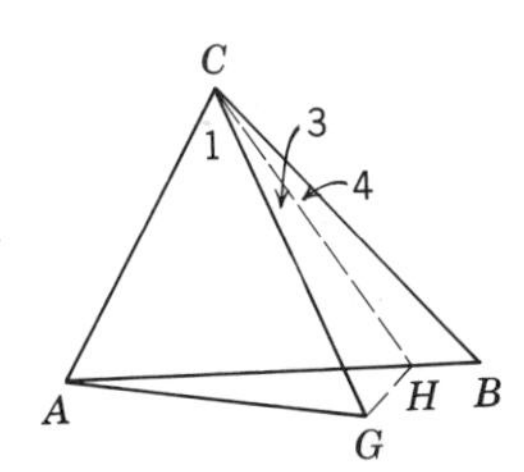

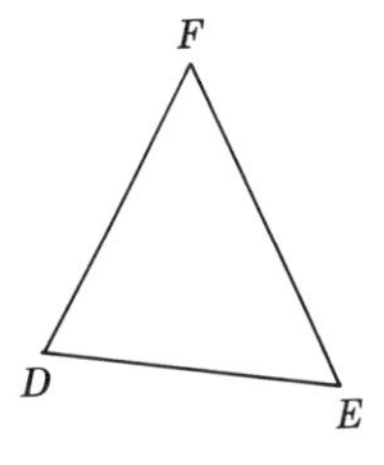

Construction. Construct $\angle 1 = \angle F$ and $GC = EF$. Construct CH, the bisector of $\angle GCB$ and draw segment GH.

Proof:

STATEMENTS	REASONS
1. $AC = DF$; $\angle C > \angle F$.	1. Given.
2. $\angle 1 = \angle F$; $GC = EF$; GC in interior of $\angle C$.	2. By construction.
3. $\triangle ACG \cong \triangle DFE$.	3. SAS = SAS.
4. $AG = DE$.	4. cpcte.
5. CH bisects $\angle GCB$.	5. By construction.
6. $\angle 3 = \angle 4$.	6. Why?
7. $CH = CH$.	7. Reflexive law.
8. $EF = BC$.	8. Given.
9. $BC = GC$.	9. Substitution.
10. $\triangle GCH \cong \triangle BCH$.	10. SAS = SAS.
11. $GH = BH$.	11. cpcte.
12. $AH + HG > AG$.	12. Postulate 7.3.
13. $AH + HB = AB > AG$.	13. Why?
14. $AB > DE$.	14. Substitution.

CASE II (point G falls on $\triangle ABC$)

Given: $AC = DF$.
$BC = EF$.
$\angle C > \angle F$.

Prove: $AB > DE$.

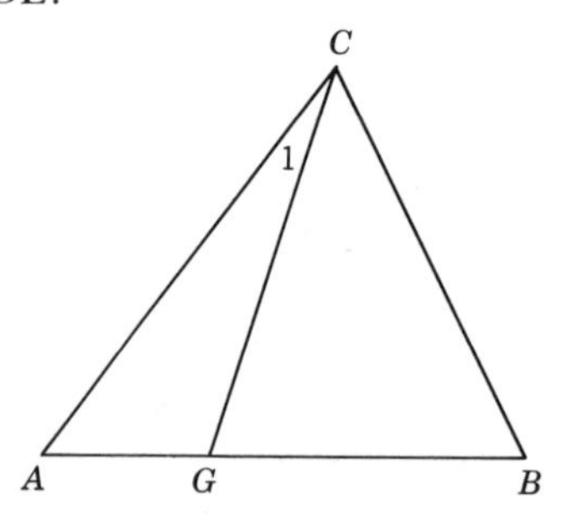

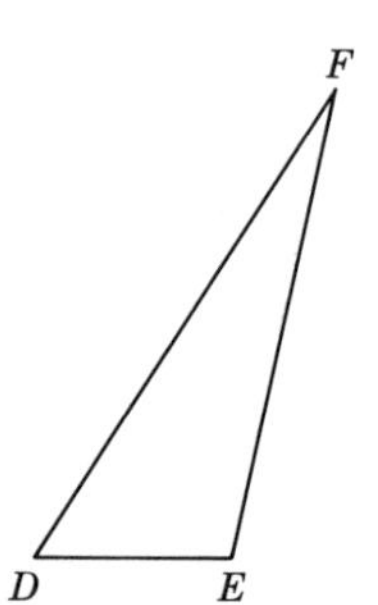

Construction. Construct $\angle 1 = \angle F$ and $GC = EF$.

The proof of Case II is left as an exercise.

CASE III (point G falls inside $\triangle ABC$)

Given: $AC = DF$.
$BC = EF$.
$\angle C > \angle F$.

Prove: $AB > DE$.

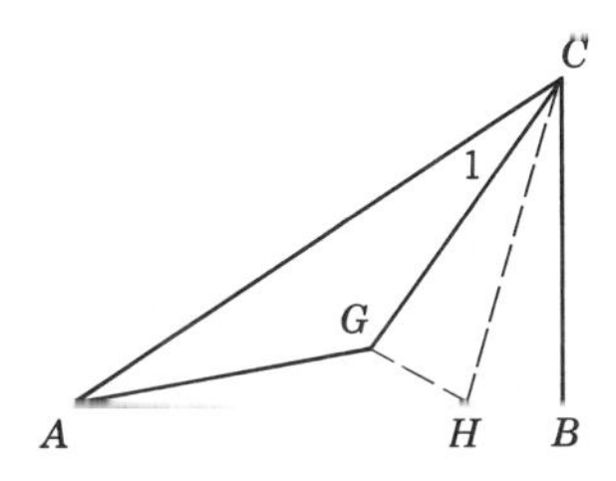

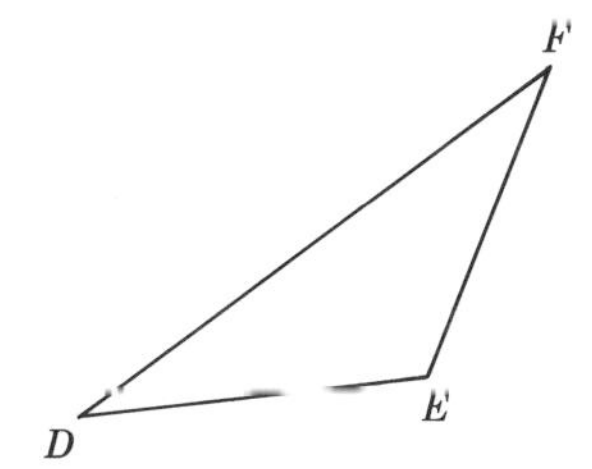

Construction. Construct $\angle 1 = \angle F$ and $GC = EF$. Also construct CH, the bisector of $\angle GCB$ and draw segment GH.

The proof of Case III is left as an exercise.

Theorem 7.12

If two sides of one triangle are equal to two sides of a second triangle and the third side of the first is greater than the third side of the second, then the angle opposite the third side of the first triangle is greater than the angle opposite the third side of the second triangle.

Given: $AC = DF$.
$CB = FE$.
$AB > DE$.

Prove: $\angle C > \angle F$.

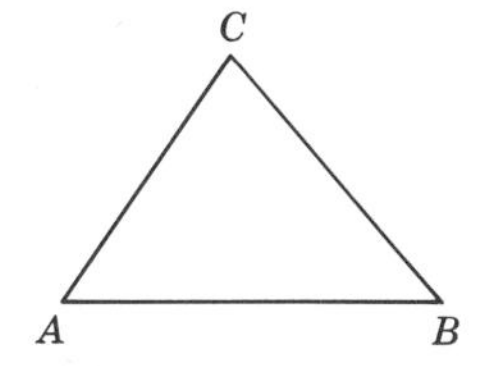

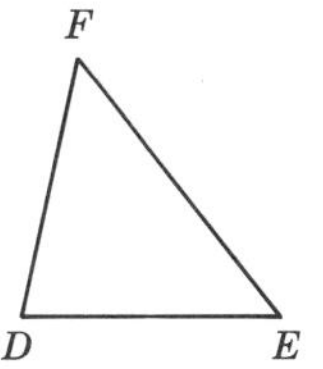

Plan: The proof is indirect. We will assume $\angle C = \angle F$, and $\angle C < \angle F$, and show that each assumption leads to a contradiction.

Proof:

STATEMENTS	REASONS
1. $\angle C = \angle F$.	1. By assumption.
2. $AC = DF$; $CB = FE$.	2. Given.

STATEMENTS	REASONS
3. $\triangle ABC \cong \triangle DEF$.	3. SAS = SAS.
4. $AB = DE$.	4. cpcte.
5. $AB > DE$.	5. Given.
6. Statement 1 is false.	6. It leads to a contradiction (statements 4 and 5).
7. $\angle C < \angle F$.	7. By assumption.
8. $AB < DE$.	8. Theorem 7.11.
9. Statement 7 is false.	9. It leads to a contradiction (statements 5 and 8).
10. $\angle C = \angle F$, $\angle C < \angle F$ or $\angle C > \angle F$.	10. Postulate 7.1.
11. $\angle C > \angle F$.	11. Only remaining possibility.

Theorem 7.13

In the same circle or in equal circles, the greater of two central angles will intercept the greater arc.

Given: $\angle 1 > \angle 2$ and $\angle 1$ and $\angle 2$ are central angles.

Prove: $\overset{\frown}{AB} > \overset{\frown}{CD}$.

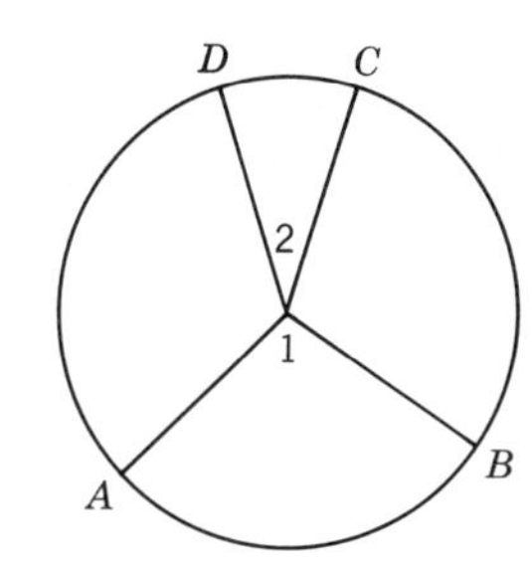

Proof:

Since $\angle 1 = \overset{\frown}{AB}$ and $\angle 2 = \overset{\frown}{CD}$ and $\angle 1 > \angle 2$, by substitution $\overset{\frown}{AB} > \overset{\frown}{CD}$.

The proof of this theorem for the case involving equal circles is left as an exercise.

Theorem 7.14

In the same circle or in equal circles, the greater of two arcs will be intercepted by the greater of the central angles.

The proof of this theorem is left for an exercise.

EXERCISES

1. Complete the proof of Theorem 7.13.
2. Prove Theorem 7.14.
3. Given the figure for exercise 3, prove that $AB > AE$.

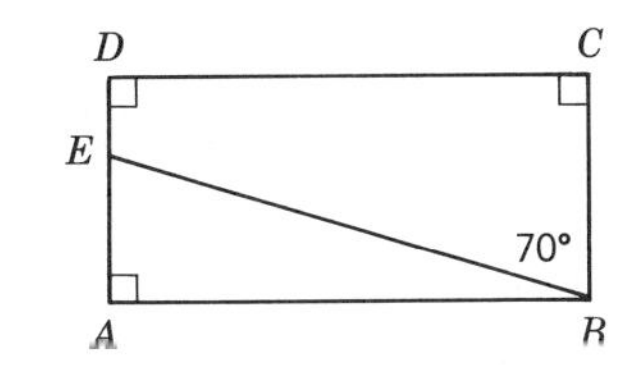

Ex. 3

4. Given the figure for exercise 4, where PA is tangent to the circle, prove $OA > PA$.

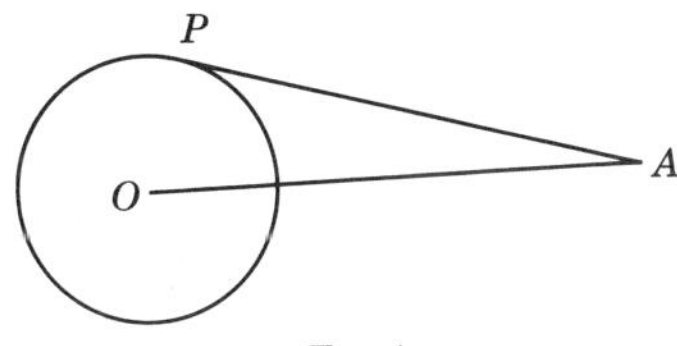

Ex. 4

5. *Prove:* the greater of two inscribed angles will intercept the greater arc.
6. $AD = DB$, $\angle A = \angle 1$ and $\angle 1 < 90°$. *Prove:* $CD < CB$.

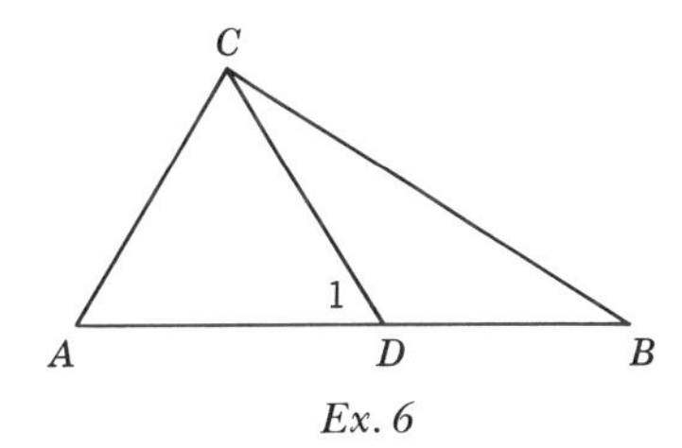

Ex. 6

7. *Prove:* the shortest distance from a point to a line is the length of a perpendicular from the point to the line.
8. Prove case II of Theorem 7.11.
9. Prove case III of Theorem 7.11.

Theorem 7.15

In the same circle or equal circles, the greater of two chords intercepts the greater minor arc.

Given: $AB > CD$.

Prove: $\overset{\frown}{AB} > \overset{\frown}{CD}$.

Construction. Draw radii OA, OB, OC, and OD.

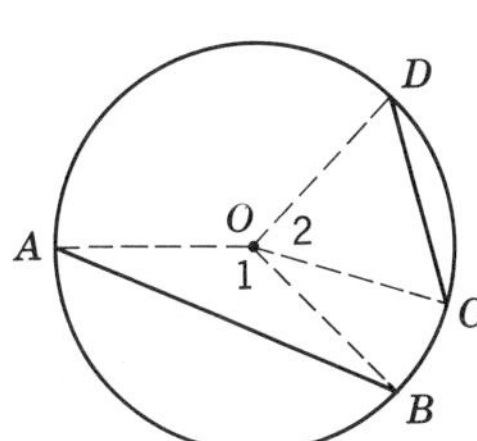

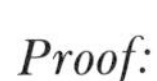

Proof:

STATEMENTS	REASONS
1. $AB > CD$.	1. Given.
2. $OA = OB = OC = OD$.	2. Why?

STATEMENTS	REASONS
3. $\angle 1 > \angle 2$.	3. Theorem 7.12.
4. $\overset{\frown}{AB} > \overset{\frown}{CD}$.	4. Theorem 7.13.

The case for equal circles is left as an exercise.

Theorem 7.16

In the same circle or equal circles, the greater of two minor arcs has the greater chord.

The proof of Theorem 7.16 is left as an exercise.

Theorem 7.17

If two unequal chords form an inscribed angle within a circle, the shorter chord is the farther from the center of the circle.

Given: $AB > AC$; $OD \perp AB$; $OE \perp AC$.

Prove: $OE > OD$.

Construction. Draw segment DE.

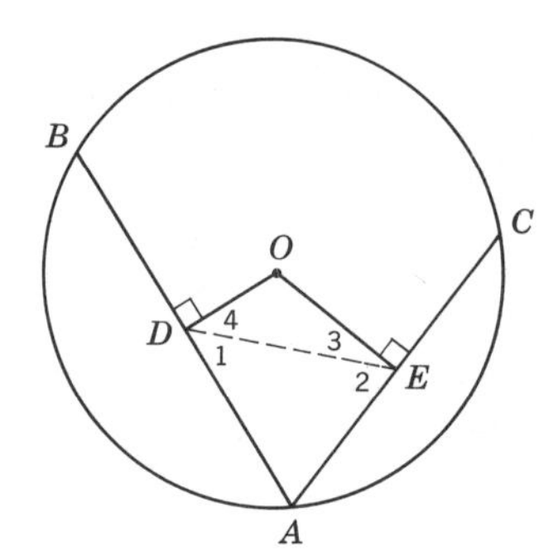

Proof:

STATEMENTS	REASONS
1. $AB > AC$; $OD \perp AB$; $OE \perp AC$.	1. Given.
2. OD bisects AB; OE bisects AC.	2. Why?
3. $\angle OEA = \angle ODA = 90°$.	3. Why?
4. $AD > AE$.	4. Why?
5. $\angle 2 > \angle 1$.	5. Theorem 7.9.
6. $\angle 3 < \angle 4$ or $\angle 4 > \angle 3$.	6. Theorem 7.3.
7. $OE > OD$.	7. Theorem 7.10.

Theorem 7.18

In the same circle or equal circles, the shorter of the two chords is the greater distance from the center of the circle.

The proof of Theorem 7.18 is left as an exercise.

EXERCISES

1. Complete the proof of Theorem 7.15.
2. Prove Theorem 7.16.
3. Prove Theorem 7.18.
4. *Prove:* if two unequal chords form an inscribed angle within a circle, the chord that is farther from the center is the smaller.

Review Test

Classify the following statements as true or false.

1. If a does not equal b, then $a < b$.
2. The commutative law of addition states that $a+b=b+a$.
3. If unequal quantities are added to unequal quantities, the sums are unequal quantities.
4. If $a > b$ and $b > c$, then $a > c$.
5. If $a > b$ and $c < d$, then $ac < bc$.
6. If a base angle of an isosceles triangle is less than 60°, the base is the longest side of the triangle.
7. The sum of the exterior angles of a triangle is greater than the sum of its interior angles.
8. The hypotenuse is greater than either arm in a right triangle.
9. The larger of two chords in a circle is nearer the center of the circle.
10. The greater of two arcs in a circle has the greater chord.
11. The greatest angle of an isosceles triangle must be the vertex angle.
12. If $a < b$, then $ac < bc$.
13. If $a < b$, then $a-c < b-c$.
14. The longest side of $\triangle ABC$ is AB.

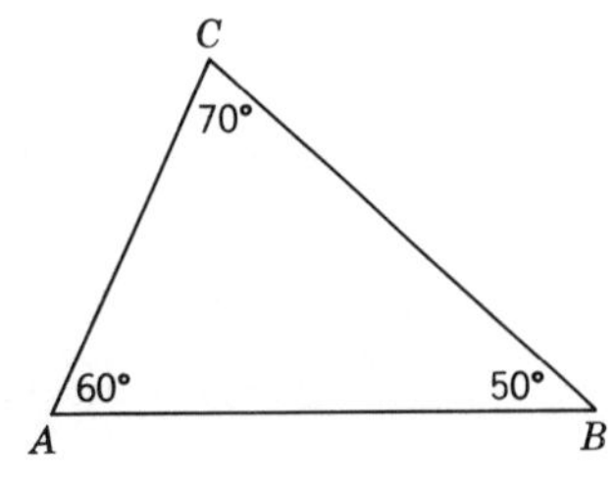

Ex. 14

15. A chord of a circle cannot be greater than a diameter of the circle.
16. The sum of the interior angles of a pentagon (five-sided figure) is greater than the sum of its exterior angles.
17. If unequal quantities are multiplied by equal quantities, the results are unequal quantities and in the same order.
18. The larger of two central angles in a circle intercepts the greater arc.
19. An exterior angle in a triangle is greater than the difference between the nonadjacent interior angles.
20. If a central angle of a circle has a measure of 50°, the chord intercepted by the angle is greater than the radius of the circle.

8

Geometric Loci

A *locus* is a set of points such that (1) all points in the set satisfy a given condition, and (2) all points that satisfy the given condition are points in the set. The word locus commonly means the place or position of a set of points.

Suppose that a dog is tied to a stake with a rope. The locus or position of the boundary of the surface over which he can roam is a circle with radius equal to the length of the rope (Fig. 8.1).

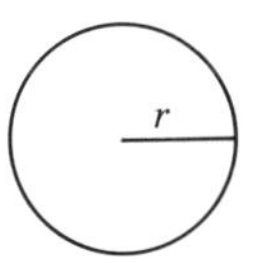

Fig. 8.1

The position or locus of the center of a car wheel as the car travels down a straight and level road is a line parallel to the road and a distance equal to the radius of the wheel above the ground (Fig. 8.2).

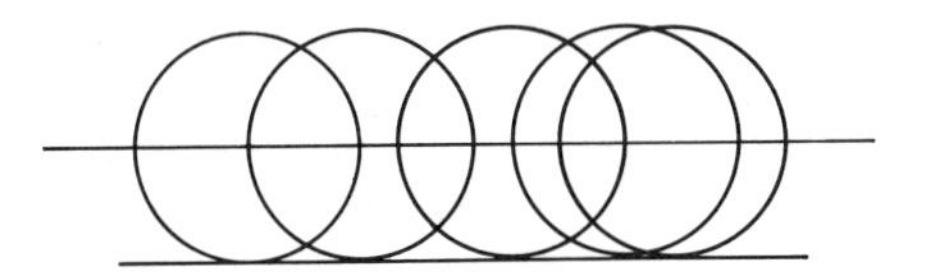

Fig. 8.2

The location or locus of all number pairs satisfying the equation $y = x$ is the graph of the equation. Each number pair that satisfies the equation represents a point on the graph and each point on the graph has a number pair for coordinates that satisfies the equation (Fig. 8.3).

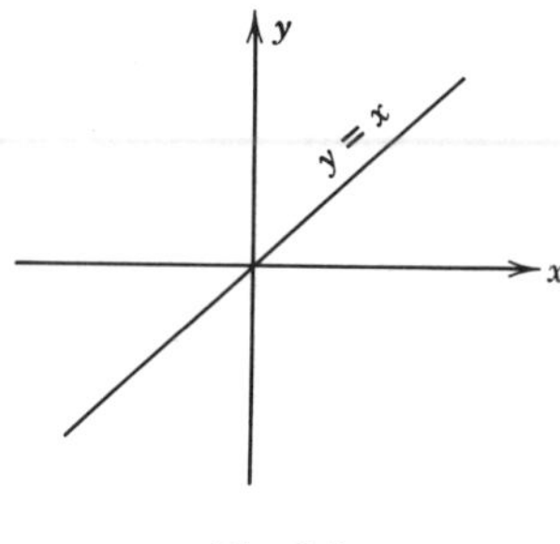

Fig. 8.3

Suppose that we wish to determine the locus of all points in a plane that are a given distance d from a fixed point in the same plane. It is easily recognized that this locus is a circle with radius d and with the fixed point as center. We now restate the definition of a circle.

Definition 8.1 A *circle* is the locus of points a given distance from a fixed point, called its *center*, and all in the same plane.

In the following theorems, we understand that all points lie in the same plane.

Theorem 8.1

The locus of a point at a given distance from a given line is a pair of lines, one on each side of the given line, parallel to the given line and at the given distance from it.

Since Theorem 8.1 is obvious, its proof has been omitted.

We now prove three locus theorems. We must remember that in each case we must prove that all points on the locus satisfy the given condition and that all points satisfying the given condition are on the locus.

Theorem 8.2

The locus of all points equidistant from two points is the perpendicular bisector of the segment joining the two points.

Part I

Given: Point P is equidistant from A and B; M is the midpoint of AB.

Prove: PM is the perpendicular bisector of AB.

Construction. Draw line segments AP and BP.

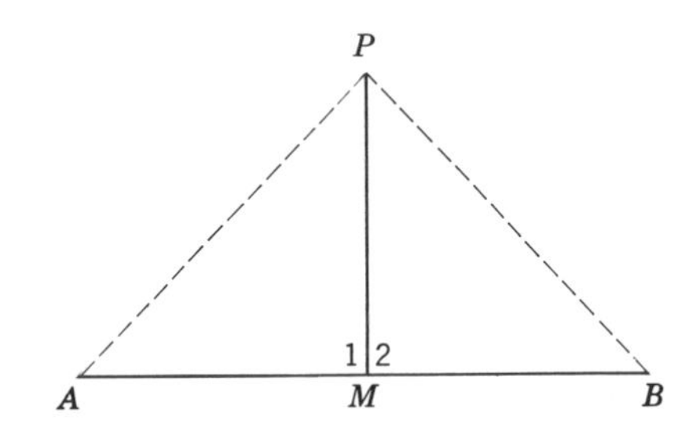

Proof:

STATEMENTS	REASONS
1. Point P is equidistant from A and B; M is midpoint of AB.	1. Given.
2. $MA = MB$.	2. Why?
3. $PM = PM$.	3. Why?
4. $AP = BP$.	4. Why?
5. $\triangle AMP \cong \triangle BMP$.	5. SSS = SSS.
6. $\angle 1 = \angle 2$.	6. cpcte.
7. $PM \perp AB$.	7. Why?
8. PM is the perpendicular bisector of AB.	8. Why?

Part II

Given: PM is the perpendicular bisector of AP.

Prove: P is equidistant from A and B.

Construction. Draw line segments AP and BP.

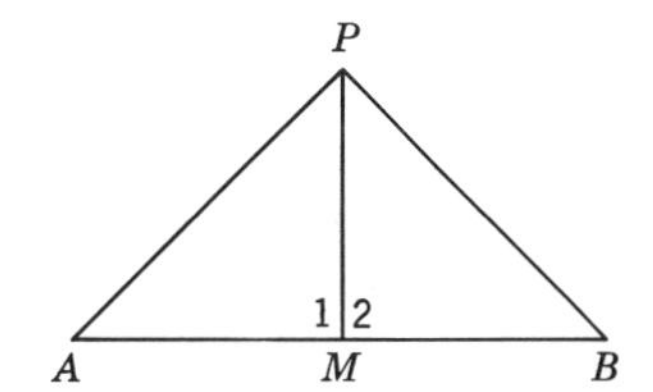

The completion of the proof of Theorem 8.2 is left as an exercise.

Theorem 8.3

The locus of all points equidistant from the sides of an angle is the angle bisector.

Part I

Given: P is equidistant from lines AB and AC.

Prove: PA is the bisector of $\angle A$.

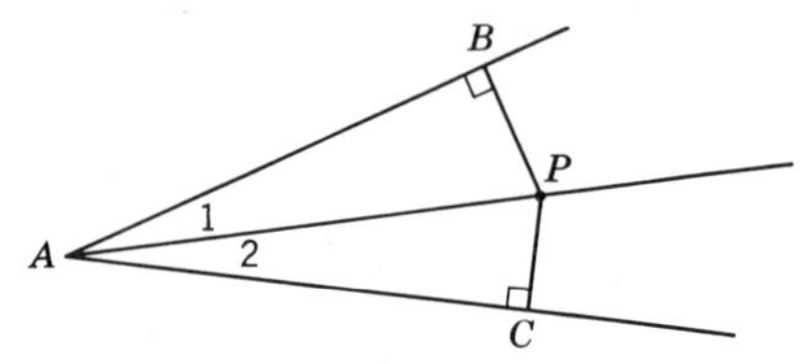

Construction. Construct perpendiculars from P to lines AB and AC.

Proof:

STATEMENTS	REASONS
1. P is equidistant from lines AB and AC.	1. Given.
2. $PB \perp AB$; $PC \perp AC$.	2. By construction.
3. $PB = PC$.	3. Why?
4. $AP = AP$.	4. Why?
5. $\triangle ABP \cong \triangle ACP$.	5. hs = hs.
6. $\angle 1 = \angle 2$.	6. cpcte.
7. PA is the bisector of $\angle A$.	7. Why?

Part II

Given: PA is the angle bisector.

Prove: P is equidistant from lines AB and AC.

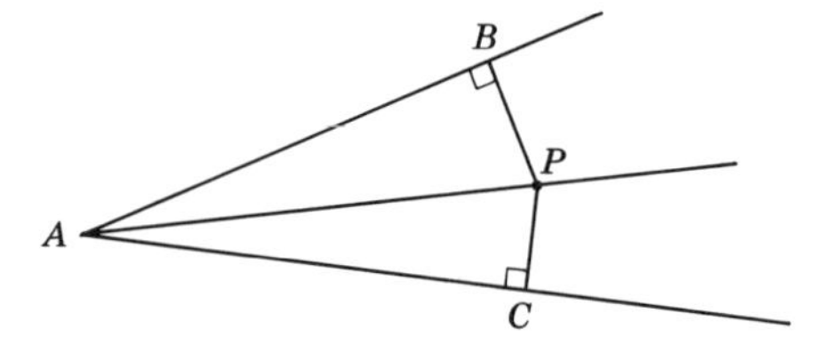

Construction. Construct perpendiculars from P to lines AB and AC.

The completion of the proof of Theorem 8.3 is left as an exercise.

Theorem 8.4

The locus of the vertex of the right angle of a right triangle with a fixed hypotenuse is a circle with the hypotenuse as diameter.

Part I

Given: $\triangle ABC$ is a right triangle with $\angle C = 90°$.
Circle O with radius OA; $OA = OB$.

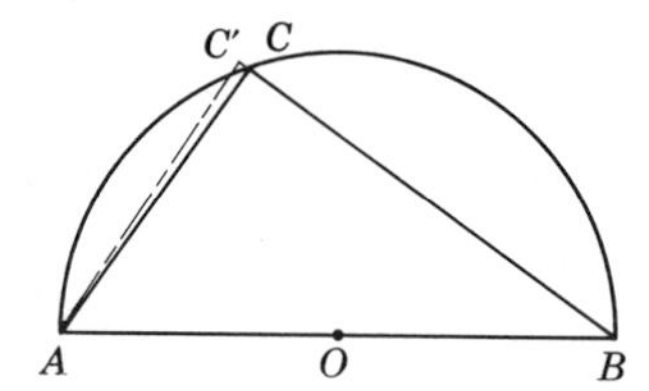

Prove: Point C is on $\widehat{AB}$.

Construction. Let the intersection of $\widehat{AB}$ and BC be point C'. Draw AC'.

Proof:

STATEMENTS	REASONS
1. C' is on $\widehat{AB}$.	1. By construction.
2. $\angle C' = 90°$.	2. Why?
3. $AC' \perp BC$.	3. Why?
4. $\angle C = 90°$.	4. Given.
5. $AC \perp BC$.	5. Why?
6. AC coincides with AC'.	6. A perpendicular to a line from a a point not on the line is unique.
7. C coincides with C'.	7. Two lines can only intersect in one point (problem 7, page 61).

Part II

Given: $\triangle ABC$;
Circle O with radius OA; $OA = OB$; C is on $\widehat{AB}$.

Prove: $\angle C$ is a right angle.

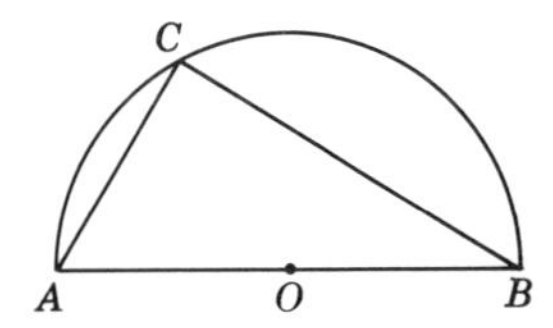

The completion of the proof of Theorem 8.4 is left as an exercise.

EXERCISES

1. Complete the proof of Theorem 8.2.
2. Complete the proof of Theorem 8.3.
3. Complete the proof of Theorem 8.4.
4. Find the locus of all points that are equidistant from two parallel lines.
5. Find the locus of the vertex angle of a triangle on a fixed base if the vertex angle is always equal to 60°.
6. Find the locus of the vertex angle of a triangle on a fixed base if the vertex angle is always equal to 45°.
7. Find the locus of all points that are equidistant from the sides of $\angle A$ and a fixed distance from point D. Discuss the possibilities.

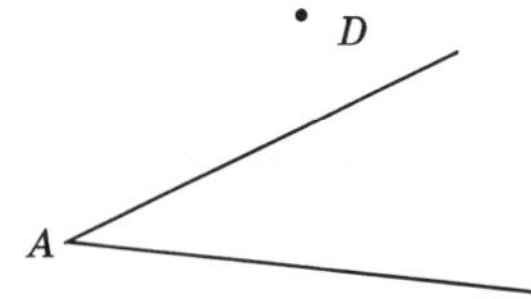

8. Find the locus of all points that are equidistant from points A and B and also at a given distance from point D. Discuss the possibilities.
9. Devise a procedure for constructing a square with area equal to that of a given rectangle. (*Hint:* recall Theorem 5.17.)
10. Devise a procedure for constructing an isosceles right triangle with a given hypotenuse.

Theorem 8.5

The perpendicular bisectors of the sides of a triangle meet at a point which is equally distant from the vertices of the triangle.

Given: $\triangle ABC$ with HD perpendicular bisector of AC and HE perpendicular bisector of AB. F is midpoint of BC.

Prove: The perpendicular bisector of BC passes through point H and H is equidistant from A, B, and C.

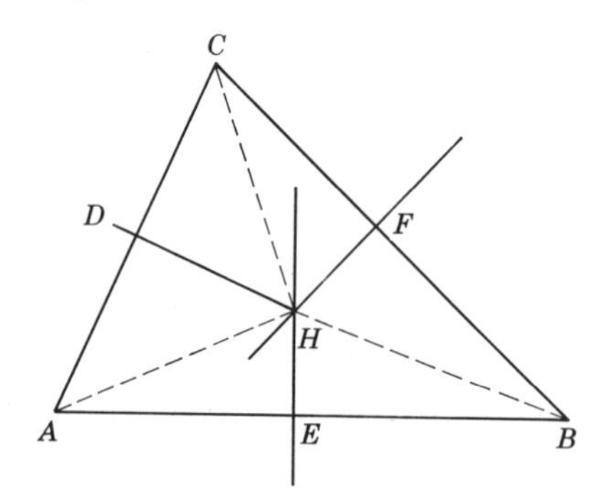

Construction. Draw segments AH, CH, and BH.

Proof:

STATEMENTS	REASONS
1. HD perpendicular bisector of AC and HE perpendicular bisector of AB.	1. Given.
2. $HC = HA$; $HA = HB$.	2. Theorem 8.2.
3. $HC = HB$.	3. Transitive law.
4. H is equidistant from A, B, and C.	4. Statements 2 and 3.
5. F is midpoint of BC.	5. Given.
6. F is on the perpendicular bisector of BC.	6. Why?

STATEMENTS	REASONS
7. H is on the perpendicular bisector of BC.	7. Statement 3 and Theorem 8.2.
8. FH is the perpendicular bisector of BC.	8. Two points determine a line.

The point of intersection of the perpendicular bisectors of the sides of a triangle is called the *circumcenter* of the triangle. Because Theorem 8.5 gives us a method of finding a point equidistant from three points, it provides a method for circumscribing a circle about a triangle. If we construct the perpendicular bisectors of two sides of $\triangle ABC$, they will meet at a point P that is

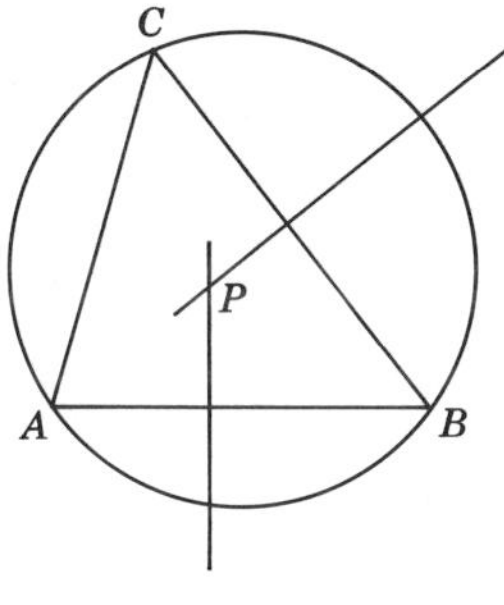

Fig. 8.4

equidistant from the vertices of the triangle. If we choose AP for the radius of the circle, the circle will pass through vertices C and B and the circle will be circumscribed about the triangle. (Fig. 8.4.)

Theorem 8.6

The angle bisectors of a triangle meet in a point which is equidistant from the sides of the triangle.

Given: $\triangle ABC$ with AP and BP bisectors of $\angle A$ and $\angle B$, respectively.

Prove: CP bisects $\angle C$ and P is equidistant from sides AB, BC, and CA.

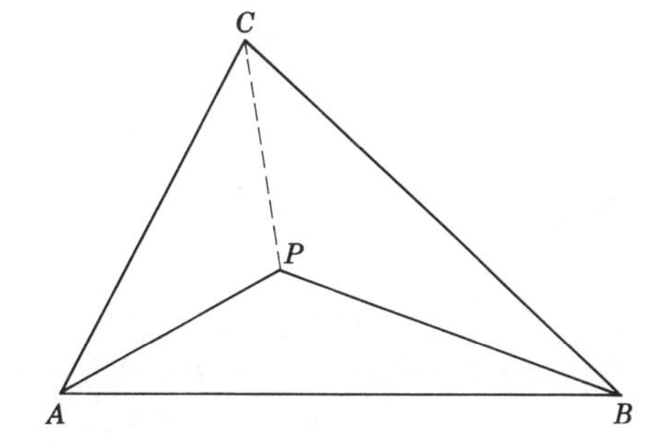

Proof:

STATEMENTS	REASONS
1. AP bisects $\angle A$; BP bisects $\angle B$.	1. Given.
2. P is equidistant from AB and AC.	2. Theorem 8.3.
3. P is equidistant from AB and BC.	3. Statement 2.
4. P is equidistant from AC and BC.	4. Why?
5. P is on the bisector of $\angle C$.	5. Theorem 8.3.
6. CP bisects $\angle C$.	6. P lies on CP, P lies on angle bisector, so CP is an angle bisector.
7. P is equidistant from sides AB, BC, and CA.	7. Statements 2 and 3.

The point of intersection of the angle bisectors of the angles of a triangle is called the *incenter* of the triangle. Theorem 8.6 gives us a method for inscribing a circle within a triangle. We construct the bisectors of two of the angles of $\triangle ABC$. These angle bisectors will meet at point P which is equidistant from sides AB, BC, and CA. We construct a perpendicular from point P to one side of the triangle, say AB, and call the point of intersection point D.

Taking point P as center and PD as radius we can construct the required circle (Fig. 8.5).

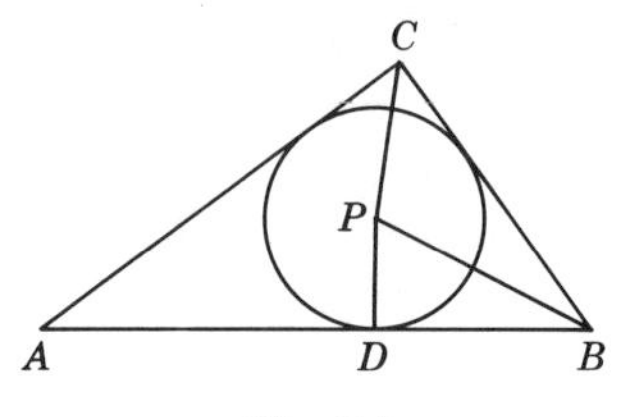

Fig. 8.5

Theorem 8.7

The altitudes of a triangle meet in a point.

Given: $\triangle ABC$ with altitudes AE, CD, and BF.

Prove: AE, CD, and BF meet in a point.

Plan: Construct $IG \| CB$, $IH \| AB$ and $GH \| AC$ and show that AE, CD, and BF are the perpendicular bisectors of the sides of $\triangle IGH$

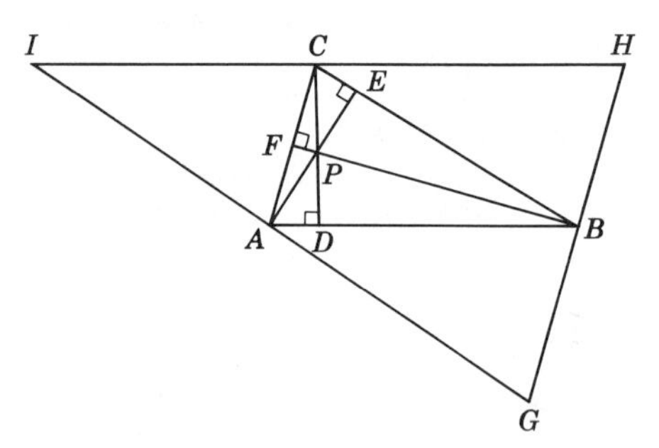

and, therefore, that they are concurrent by Theorem 8.5. It is easy to show that AE, CD, and BF are perpendicular to IG, IH, and GH, respectively. To show point A is the midpoint of segment IG consider parallelograms $ABCI$ and $AGBC$. In a similar fashion point B and point C can be shown to be midpoints.

The details of the proof of Theorem 8.7 are left as an exercise.

The point of intersection of the altitudes of a triangle is called the *orthocenter* of the triangle.

Theorem 8.8

The medians of a triangle meet in a point.

Given: $\triangle ABC$ with medians NA and MB, intersecting at point P.

Prove: CPQ is a median of $\triangle ABC$.

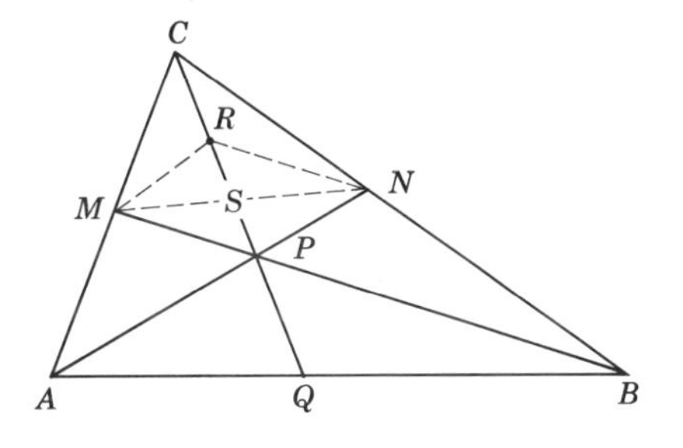

Plan: Locate the midpoint of CP and call it point R. Draw segments MR, RN, and MN. $MRNP$ is a parallelogram. Why? $MS = SN$ because the diagonals of a parallelogram bisect each other. Since $\triangle ACQ \sim \triangle MCS$ and $AC = 2MC$, $AQ = 2MS$. Similarly $QB = 2SN$ and $AQ = QB$ and, therefore, CPQ is a median of $\triangle ABC$.

The details of the proof of Theorem 8.8 are left as an exercise.

The point of intersection of the medians of a triangle is called the *centroid* of the triangle.

Theorem 8.9

The point of intersection of the medians of a triangle is 2/3 of the way from the vertex to the midpoint of the opposite side.

Given: $\triangle ABC$ with medians AN, BM and CQ.

Prove: $AP = \frac{2}{3}AN$; $BP = \frac{2}{3}BM$ and $CP = \frac{2}{3}CQ$.

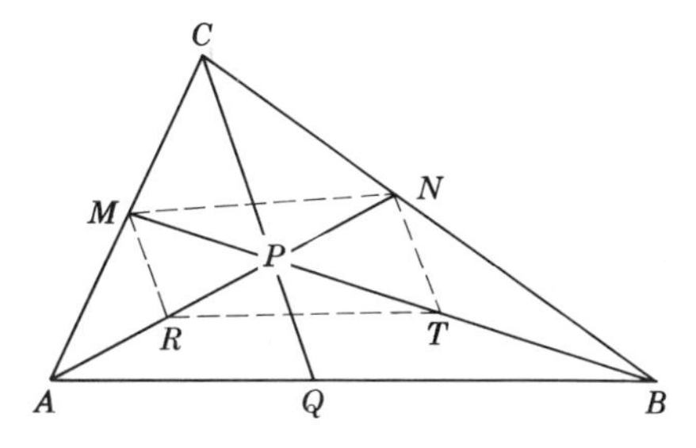

Plan: Locate the midpoints of segments AP and BP and call them R and T, respectively. Draw segments MR, RT, MN, and TN. Since $MRTN$ is a parallelogram (why?) $RP = PN$ and $TP = PM$. Why? Therefore $AR = RP = PN$ and $BT = TP = PM$ and $AP = \frac{2}{3}AN$ and $BP = \frac{2}{3}BM$. To show $CP = \frac{2}{3}CQ$, consider the midpoints of segments PC and AP and proceed with a similar argument.

The details of the proof of Theorem 8.9 are left as an exercise.

EXERCISES

1. Complete the proof of Theorem 8.7.
2. Complete the proof of Theorem 8.8.
3. Complete the proof of Theorem 8.9.
4. Cut out a large triangle from a piece of cardboard and very carefully locate its centroid. Try to balance the triangle on the point of a pencil by placing the pencil point on the centroid. What do you discover? Will the triangle balance at any other point?
5. Draw a triangle and locate its orthocenter.
6. Draw a triangle and locate its incenter.
7. Draw a triangle and locate its circumcenter.
8. Can a circle be drawn through any four points? Why?
9. Circumscribe a circle about a square.
10. Inscribe a circle within a square.
11. Can the orthocenter, incenter, circumcenter, and centroid be concurrent in any triangle? Explain.
12. If an altitude of an equilateral triangle is 4 inches, find the radius of the circumscribed circle. Find the radius of the inscribed circle.

Review Test

Classify the following statements as true or false.

1. The altitudes of a triangle meet in a point that is in the interior of the triangle.
2. The angle bisectors of the three angles of a triangle meet in a point that is in the interior of the triangle.
3. The medians of a triangle meet in a point that is in the interior of the triangle.
4. The locus of points equidistant from two points is a bisector of the line segments joining the two points.

5. The locus of points equidistant from the sides of an angle is the bisector of the angle.

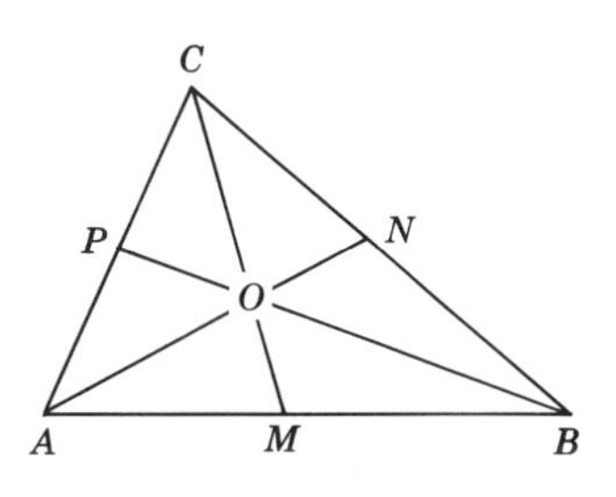

6. M, N, and P are midpoints of the three sides of $\triangle ABC$. If $OM = 2$ inches, then $ON = 2$ inches.
7. M, N, and P are midpoints of the three sides of $\triangle ABC$. If $ON = 2$ inches, then $AO = 4$ inches.
8. The locus of points that are equidistant from the sides of an angle and at a given distance from the vertex of the angle contains exactly one point.
9. The locus of points equidistant from two given parallel lines is a line, parallel to the given lines and midway between them.
10. The locus of the vertex of the right angle of a right triangle with fixed hypotenuse is a circle minus the endpoints of the hypotenuse.
11. The point of intersection of the altitudes of a triangle is called the orthocenter of the triangle.
12. The point of intersection of the medians of a triangle is called the centroid of the triangle.
13. The point of intersection of the angle bisectors of a triangle is called the circumcenter of the triangle.
14. It is always possible to construct a circle passing through three given points if the points are not collinear.
15. The locus of points equidistant from a line segment is a pair of parallel lines.

9

Regular Polygons and Circles

There are some interesting theorems that can be proved about regular polygons that are either inscribed in or circumscribed about a circle. We recall that a regular polygon is one with equal angles and equal sides. If a polygon has all of its vertices on a circle, it is said to be *inscribed* within the circle or the circle is said to be *circumscribed* about the polygon. If the sides of a polygon are all tangent to a circle, the polygon is said to be circumscribed about the circle or the circle is said to be inscribed within the polygon.

Theorem 9.1

If a circle is divided into n equal arcs, the chords of the arcs will form a regular polygon. ($n > 2$)

Given: $\overparen{AB} = \overparen{BC} = \overparen{CD} = \overparen{DE} = \overparen{EA}$.

Prove: $ABCDE$ is a regular polygon.

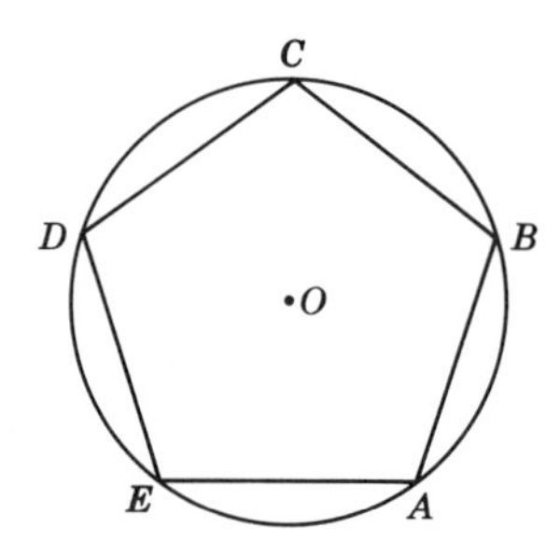

Proof: The circle is divided into n equal arcs. For the sake of argument, we shall take $n = 5$. Since $\overparen{AB}$, $\overparen{BC}$, $\overparen{CD}$, $\overparen{DE}$, and $\overparen{EA}$ are all equal arcs and in the same circle equal arcs have equal chords, then $AB = BC = CD = DE = EA$. Each angle formed by any two chords of the circle is an inscribed angle that intercepts an arc equal to 3 times the measure of a single arc. Hence, all angles of the inscribed polygon are equal. Because *ABCDE* has all sides equal and all angles equal, it is a regular polygon.

The preceding argument can be generalized to any number of equal arcs.

Theorem 9.2

If a circle is divided into n equal arcs and tangents are drawn to the circle at the endpoints of these arcs, the figure formed by the tangents will be a regular polygon. ($n > 2$)

Given: $\overset{\frown}{AB} = \overset{\frown}{BC} = \overset{\frown}{CD} = \overset{\frown}{DE} = \overset{\frown}{EA}$
MN, NP, PQ, QR and RM are tangents to the circle.

Prove: $MNPQR$ is regular polygon.

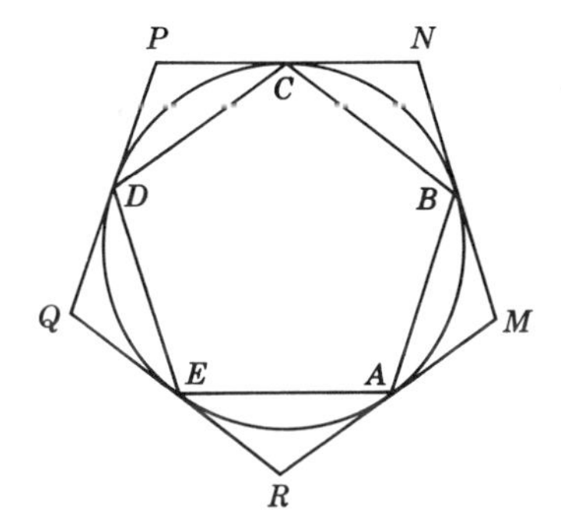

Proof: The circle is divided into n equal arcs. For argument we take $n = 5$. Since all of the arcs are equal, all of the chords are equal. Because tangents drawn to a circle from a point outside the circle are equal, triangles AMB, BNC, CPD, DQE and ERA are isosceles triangles. Angles formed by a tangent and a chord have measure equal to one-half their intercepted arc and since all five arcs are equal, the base angles of the above triangles are equal. The five above triangles are congruent by ASA = ASA. Angles M, N, P, Q and R are equal because they are corresponding parts of congruent triangles. Also, $MN = NP = PQ = QR = RM$. Why?

Since $MNPQR$ has equal sides and equal angles it is a regular polygon.

The preceding argument can be generalized to any number of equal arcs.

Theorem 9.3

A circle can be inscribed in any regular polygon.

We will give a procedure for inscribing a circle within a regular polygon and thereby demonstrate that it can be done. Given the regular polygon *ABCDE*, construct the bisectors of two of the angles, say $\angle A$ and $\angle B$. Call their point of intersection point P. Construct a perpendicular PR from point P to one

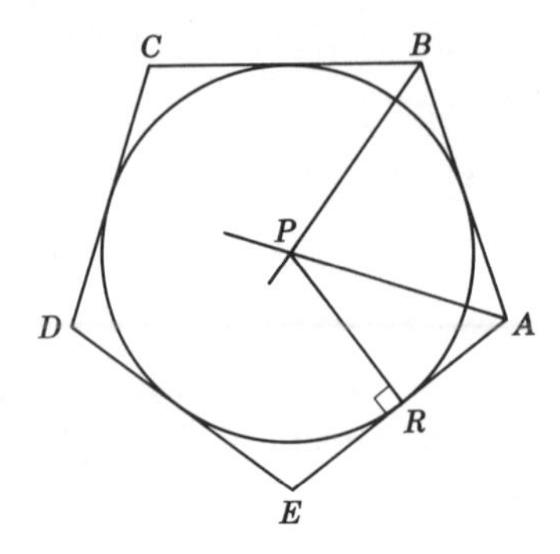

side of the polygon. Take point P as center and PR as radius and draw circle P. Circle P is inscribed within regular polygon $ABCDE$.

The proof of this construction is left as an exercise.

Theorem 9.4

A circle can be circumscribed about any regular polygon.

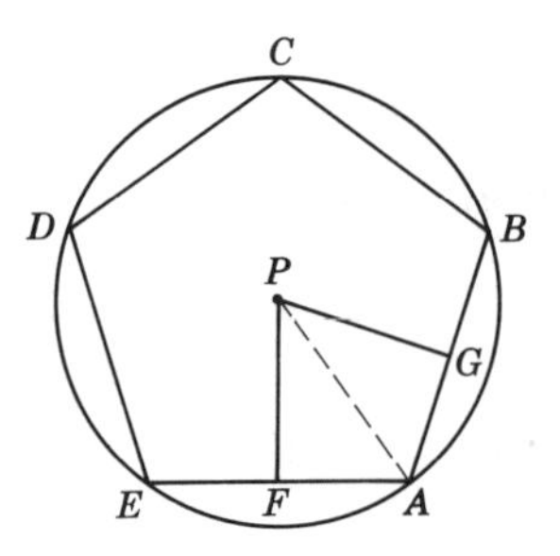

As in the preceding theorem, we demonstrate a procedure for the construction. Construct the perpendicular bisectors FP and GP of two sides of the given regular polygon $ABCDE$. Call their intersection point P. Using point P as center and segment PA as radius, draw circle P. Circle P is circumscribed about regular polygon $ABCDE$.

The proof of this construction is left as an exercise.

Definition 9.1 The *center of a regular polygon* is the center of the circle inscribed within the polygon.

Definition 9.2 The *radius* of a regular polygon is a line segment drawn from the center of the polygon to one of its vertices.

Definition 9.3 A *central angle of a regular polygon* is the angle formed by radii drawn to two consecutive vertices.

Theorem 9.5

The center of a circle that is circumscribed about a regular polygon is the center of the circle that is inscribed within the regular polygon.

Given: Circle O is circumscribed about regular polygon $ABCDE$.

Prove: O is the center of the circle inscribed in regular polygon $ABCDE$.

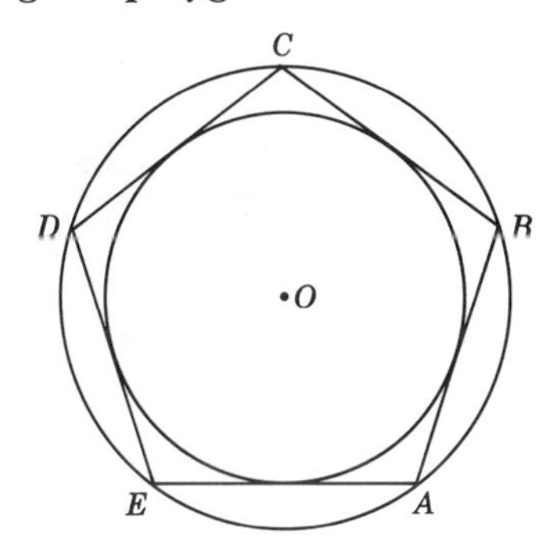

The proof of Theorem 9.5 is left as an exercise.

EXERCISES

1. Prove the construction given for Theorem 9.3.
2. Prove the construction given for Theorem 9.4.
3. Prove Theorem 9.5. (*Hint:* equal chords are equidistant from the center of a circle.)
4. Construct a regular hexagon and circumscribe a circle about it.
5. Construct a regular octagon and inscribe a circle within it.
6. Inscribe a circle within a square.
7. Can a circle be inscribed in a rhombus that is not a square?
8. In a given circle, inscribe an equilateral triangle.
9. Construct a regular polygon with a central angle of 30°.
10. Can a regular polygon with a central angle of 32° be constructed?

Definition 9.4 An *apothem* of a regular polygon is a line segment drawn from the center of the polygon and perpendicular to one of its sides.

Fig. 9.1. Karl Fredrick Gauss (1777–1855). Gauss is acknowledged to be one of the greatest mathematicians of all time. In addition to his work in geometry, he made significant contributions to the fields of arithmetic, number theory, analysis, and algebra as well as astronomy and physics. Gauss considered as one of his greatest discoveries a method for constructing a regular polygon of 17 sides. (Deutsches Museum, Munich).

Theorem 9.6

An apothem of a regular polygon bisects its respective side.

Given: $ABCDE$ is a regular polygon.
OM is an apothem.

Prove: OM bisects EA.

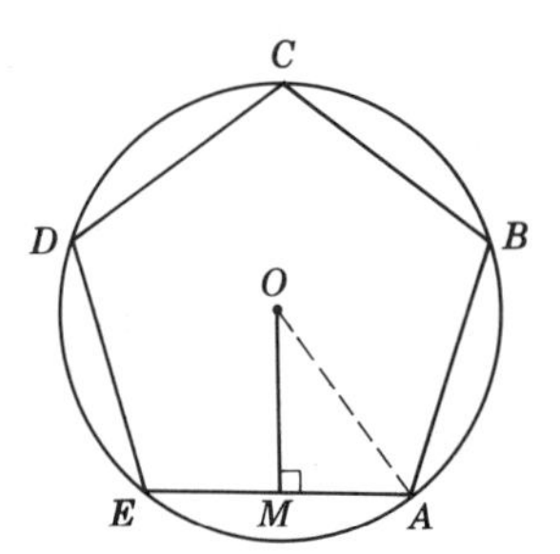

The proof of Theorem 9.6 is left as an exercise.

Theorem 9.7

A central angle of a regular polygon is given by the formula, $x = 2/n \cdot 180°$, where n is the number of sides of the polygon.

Given: $ABCDE$ is a regular polygon.

Prove: $x = (2/n) \cdot 180°$.

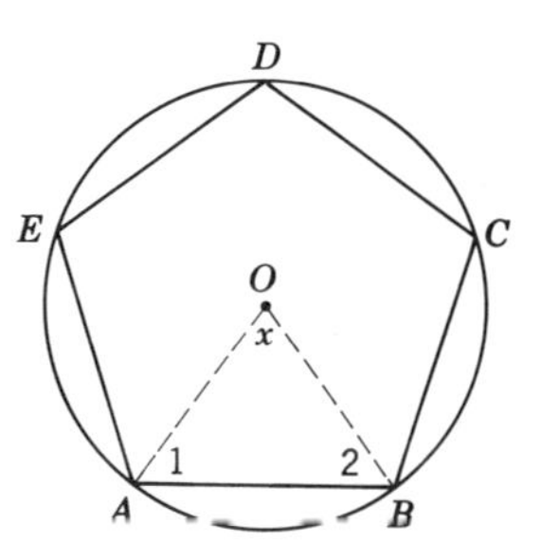

Proof: Each angle of regular polygon $ABCDE$ equals $(n-2)180/n$. OA and OB bisect angles A and B. Why? The sum of $\angle x$, $\angle 1$ and $\angle 2$ equals 180°. The rest of the proof is left as an exercise.

In exercise 11 on page 173, we show that all apothems of the same regular polygon are equal. Because this is true, it is common to say that any apothem of a regular polygon is *the* apothem of that polygon.

Theorem 9.8

The area of a regular polygon is equal to the product of one-half its apothem and its perimeter.

Given: $ABCDE$ is a regular polygon.
OM is an apothem.

Prove: $A(ABCDE) = \frac{1}{2}(OM)(AB + BC + CD + DE + EA)$.

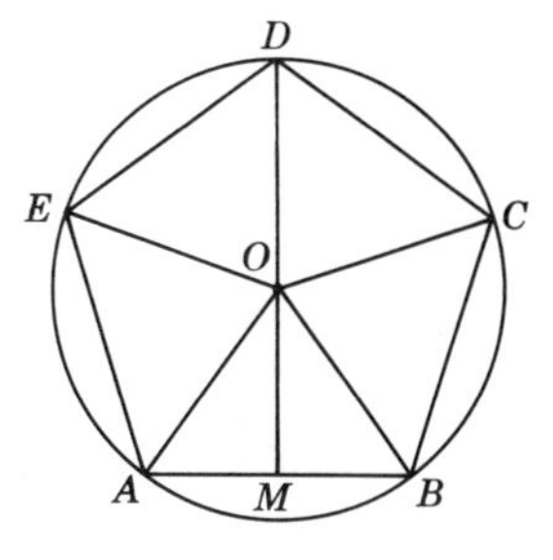

We find the areas of triangles ABO, BCO, CDO, DEO, and EAO and add them together to find the area of $ABCDE$.

The details of the proof are left as an exercise.

We recall a definition of Chapter 4.

Definition 4.2 The *circumference* of a circle is the distance about a circle. The circumference of a circle is given by the formula, $C = 2\pi r$ where r is the radius of the circle.

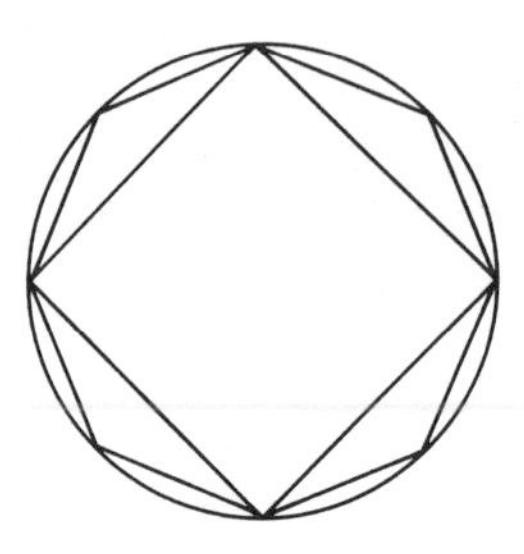

Fig. 9.2

Suppose that we are given a circle and inscribe a square within it (see Fig. 9.2). The area of the square will be an approximation to the area of the circumscribed circle. Now suppose we bisect each arc associated with the sides of the square and form a regular octagon using the four midpoints of the arcs and the vertices of the square. The area of this octagon will be a much better approximation to the area of the circle. Why? We could then repeat the process again and construct a regular polygon of 16 sides. The area of this polygon would be a very good approximation to the area of the circumscribed circle. In general, the more sides the regular polygon has, the better the approximation will be.

The area of this regular polygon is at each stage equal to $\frac{1}{2}ap$, where a is the apothem, and p the perimeter of the polygon.

In this limiting process, as the number of sides of the regular polygon increases, the apothem approaches the length of a radius. The perimeter of the polygon also approaches the circumference of the circle.

The area therefore approaches $\frac{1}{2}rC = \frac{1}{2}r(2\pi r) = \pi r^2$. This proves the following theorem, which we conjectured in Chapter 4.

Theorem 9.9

The area of a circle is given by the formula, $A = \pi r^2$ where π is approximately 3.14 and r is the radius of the circle.

Definition 9.5 A *sector* of a circle is the set of points between two radii and their intercepted arc.

Area OAB is a sector of circle O (Fig. 9.3).

Axiom 9.1 The area of a sector of a circle divided by the area of the circle is equal to the measure of its central angle divided by 360°.

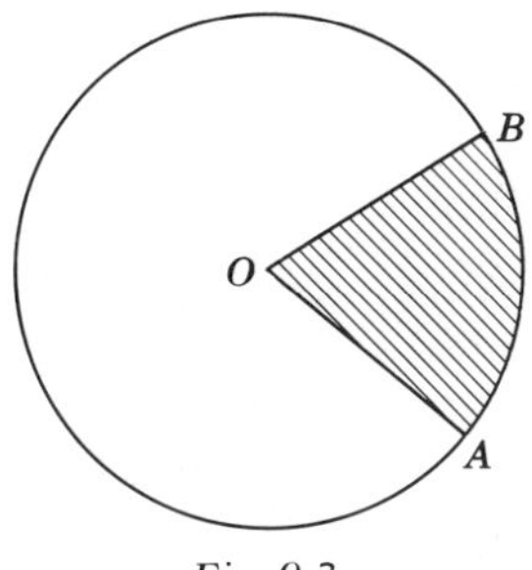

Fig. 9.3

EXERCISES

1. Prove Theorem 9.6.
2. Complete the proof of Theorem 9.7.
3. Find the measure of the central angle of a regular octagon, decagon.
4. Complete the proof of Theorem 9.8.
5. $\triangle ABC$ is an equilateral triangle inscribed in circle O with radius 8 inches. Find the area of $\triangle ABC$.

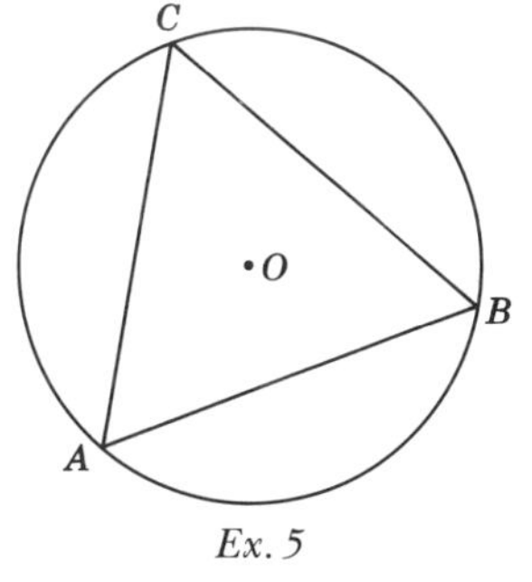

Ex. 5

6. Find the area of a regular decagon with side 10 inches long if its apothem measures 15 inches.
7. Two circles are called concentric circles if they have the same point as center. Prove that the area between two concentric circles is given by the formula $A = \pi(R^2 - r^2)$ if the two circles have radii of R and r, respectively.
8. A football stadium has a circular track. The longest distance that a person can run in a straight line and still remain on the track is 100 yards. Find the area of the track.
9. Find the area of a sector of a circle with a central angle of 50° if the circle has a radius of 7 inches.
10. Find the radius of a circle if its perimeter is equal to its area.
11. *Prove:* each apothem of a regular polygon has equal measure.
12. *Prove:* if two regular polygons are similar, the ratio of their perimeters is equal to the ratio of their apothems.

*Fig. 9.4 Aristotle (384–322*B.C.*) Aristotle, a Greek philosopher and logician, was a student of Plato and in turn a tutor of Alexander the Great. One of his greatest works was the Organum, a set of six treatises on logic. The logic of Aristotle forms the basis of the formal logic studied today. This representation is from the Cathedral of Chartres, France.* (New York Public Library Picture Collection.)

Review Test

Classify the following statements as true or false.

1. A rhombus is a regular polygon.
2. If a regular polygon is inscribed in a circle, the polygon will divide the circle into equal arcs.
3. Any rectangle can be inscribed within a circle.
4. The center of an inscribed polygon coincides with the center of its circumscribed circle.
5. An apothem is the radius of an inscribed regular polygon.
6. The perimeter of an inscribed polygon is equal to one-half the apothem multiplied by the area of the polygon.

7. The circumference of a circle is the distance about the circle.
8. A central angle of a regular pentagon is 72°.
9. The area of a circle with diameter of 14 feet is 49 square feet.
10. The circumference of a circle with radius of 6 inches is 12π inches.
11. A sector of a circle is a triangle.
12. If a polygon is inscribed in a circle, it is a regular polygon.
13. If a regular 12-sided figure is inscribed in a circle, each of the 12 arcs formed will equal 30°.
14. The circumference of a circle with radius of 1 inch is equal to the circle's area.
15. In the same circle sectors with equal central angles always have equal areas.

If it was so, it might be:
if it were so, it would be.
But, as it isn't, it ain't—
that's logic.

Lewis Carroll (1832–1898)

10 *Mathematical Logic*

There is an old saying that "Life is full of surprises." This probably is true for puppies, baby ducks, and the like, but is not entirely true for humans. Of course, unexpected events do occur, but by and large much of our life is expected and predictable. How often do we find ourselves saying, "I should have known." "Yep, just what I figured." or "Of course, what did you expect?" Human beings need not be surprised all the time, because humans are able to reason; we are able to take facts, observations, or situations, and on the basis of past experience or "common sense" come up with certain expected results. When the events occur as predicted, we are not surprised.

In this section, we are going to look at some of the principles behind the "common sense" of forming conclusions. There may be times when common sense does not seem to make any sense at all, and we will have to ask the instructor for help. Go right ahead and ask. When we do see what is going on, logic can really be fun.

Probably no one in this class has ever met a Martian. Except for their pale green color and a few quirks in their language, they are really quite like us. Their language is basically English, although it contains a few words that we do not understand. Let us listen in on a conversation. The tall one is speaking:

"You know full well that if the fribbit goes pflatz, then we will have to repair the tagget. Right?"

"Right!" says the little one.

"And you heard the fribbit this morning, didn't you? Right in the middle of breakfast. Pflatz! Right?"

"Right." The little one is a bit sheepish, now.

"Then obviously, we had better . . . "

The rest of the conversation is lost as the air-lock door slides shut behind them.

Read the conversation over again. Can you finish the last sentence? Does the tagget need repair?

We have no idea what a fribbit is, nor how to repair a tagget, yet we are able to come to the conclusion that the tagget needs to be fixed. How is this? This conclusion is sound, and we have arrived at it without having the faintest idea of what the Martians are talking about.

If we accept the two statements:

"If the fribbit goes pflatz, then we must repair the tagget." and
"The fribbit went pflatz."

then, regardless of what these things mean, we are forced to conclude that

"We must repair the tagget."

whatever *that* is.

Since the meanings are unknown, the conclusion must be determined solely by the structure of the sentences:

If p happens, then q occurs.
p happens.
———————————
therefore q occurs.

We can write this structure more briefly as:

If p, then q
p
————
therefore q

and even more compactly as:

$$\begin{array}{l} p \rightarrow q \\ \underline{p\quad\quad} \\ \therefore q \end{array}$$

where "$\rightarrow$" means "if . . . , then . . . " and "$\therefore$" means "therefore."

Whatever the notation, the form of this argument has been recognized as valid logic ever since Aristotle. The form has an ancient name: *modus ponens*, or the *law of detachment*.

Now let us reverse the situation. Certain common everyday earth words are complete gibberish to a Martian. A Martian overhears the following earth conversation. The words he cannot understand are printed in boldface.

"If I have any **children**," says one earth man, "then I must certainly be a **parent**. Right?"
"Yes, I'll go along with that." says the other.
"And I am a **parent**, aren't I?"
"Yep."
"Well then, obviously . . ."

and the conversation is lost as the earthlings walk away.

The Martian, remember, has no idea of what these sentences refer to, since he has no understanding of the words "parent" and "children." Is there something in the structure of the sentences that will lead him to a conclusion? If so, what would the conclusion be? Do not skip over these two questions; pretend *you* are the Martian, and answer them.

If we could tune in on the Martian's thoughts, we might hear this:

> "I guess if an earth man has **children**, whatever they are, then he has to be one of these **parent** things. That's what the earthman said; I don't understand it, but I'll accept it. And he *is* a **parent**, or so he says. I wonder if that means he has **children**? Maybe, and maybe not. Is having **children** the *only* way he can be a **parent**? For all I know, he could be a **parent** if he had a couple of well-repaired taggets. I guess I can't come to a conclusion."

The structure of these statements is:

If I have children, then I'm a parent.
I am a parent.
___
therefore??????????

This structure warrants *no* conclusion. It is very tempting to conclude, "therefore I have children." and this conclusion is *true*, but unfortunately it is *invalid*, because it does not follow logically from the statements that are given. The Martian is correct; no valid conclusion is possible.

This is the essence of logic. Sentences in certain forms force certain conclusions, and it is the *form* of the statements and not their *contents* which determines when a conclusion is warranted. We Martians and earthlings may not be able to understand what the other is talking about, but we still should be able to follow each other's reasoning.

EXERCISES

1. Distinguish between a *true* conclusion, and a *valid* conclusion. Is the conclusion "We must repair the tagget" a valid conclusion? Is it a true conclusion?
2. Is a valid conclusion necessarily true? Is a true conclusion necessarily valid?
3. Examine the following arguments:

   If I have children, then I am a parent.
   I am a parent.
   ___
   therefore, I have children.

   If I have measles, then I am sick.
   I am sick.
   ___
   therefore, I have measles.

Do these arguments have the same form? (Change "children" to "measles," and "a parent" to "sick". Does that change the first argument into the second?) Is the second argument valid? Is the first?

4. Examine this argument:

   If my dog has fleas, then my dog is a Republican.
   My dog has fleas.
   ____
   therefore, my dog is a Republican.

   Is this argument valid? (If you aren't sure, pretend you are a Martian who has never heard of dogs, fleas, or Republicans.) Is the conclusion true?

5. Based on your answers to questions 3 and 4, go back and answer question 2 again.

Give a valid conclusion to each of these arguments, if possible. If no conclusion is possible, write "*No Conclusion.*"

6. If it is Friday, tomorrow is a holiday.
   It is Friday.
   ____

7. If it is raining, then it is cloudy.
   It is cloudy.
   ____

8. If I study hard, then I will pass geometry.
   I study very hard.
   ____

9. The mail is late if it is snowing.
   It is snowing.
   ____

10. Look up the meaning of the Latin, "modus ponens." Your instructor can provide some references.

11. If the reasoning is valid, and the premises are true, must the conclusion be true? If the premises are false, must the conclusion be false?

We now realize that the form of an argument determines its validity, and we have seen one form of valid argument, *modus ponens*. We will now examine others. Consider this example:

If I add 3 and 4, my answer is 7.
My answer is *not* 7.
____

What conclusion follows here? Obviously, it is "I did *not* add 3 and 4," because if I *had* added 3 and 4, the result would have been 7 as promised by the first statement, contradicting the second statement. The form of this argument is:

If p happens, then q occurs
q does *not* occur.
____
Therefore, p did *not* happen.

Fig. 10.1 David Hilbert (1862–1943). Hilbert's work was primarily in the foundations of geometry. His book Grundlagen der Geometrie (Foundations of Geometry) *examined the postulates of geometry and presented a more complete list of postulates that eliminated many of the flaws in logic found in Euclid. This careful scrutinizing of the foundations of 2000-year-old Euclidean geometry led to a careful analysis of the structure of many other branches of mathematics.* (Historical Picture Service/Chicago).

More briefly, this can be written as:

$$\begin{array}{l} p \rightarrow q \\ \underline{\sim q \qquad\qquad} \\ \therefore \sim p \end{array}$$

Where "$\sim$" means "it is not so that . . .". This form, too, has an ancient name, *modus tollens*.

EXERCISES

Give a valid conclusion to each of these arguments, if possible. If no conclusion is possible, write "*No Conclusion.*"

1. If I don't get a diamond, our engagement is off.
   Our engagement is *not* off.
2. If it is hot and humid, it will rain.
   It will not rain.
3. I'll play poker Saturday night, if my wife will let me.
   My wife won't let me do anything.
4. If geometry is fun, then I'm a monkey's uncle.
   I'm not a monkey's uncle.
5. If I have to do one more of these silly things, I'll go batty.
   I'll not go batty.

In an "if . . .then . . ." statement, the "if" clause is known as the *antecedent*, and the "then" clause is called the *consequent*. For example, the antecedent of "If today is Friday, then tomorrow is a holiday." is the statement, "today is Friday." The consequent is "tomorrow is a holiday."

The positions of antecedent and consequent cannot be interchanged without changing the entire meaning of the statement. "If today is Friday, then tomorrow is a holiday." is a true statement, (since Saturday is a holiday) but when antecedent and consequent are interchanged, the sentence is false: "If tomorrow is a holiday, then today is Friday."

The new statement obtained by exchanging antecedent and consequent in an "if . . . then" statement is called the *converse* of the "if . . . then . . ." statement.

A question now arises. If an "if . . . then . . ." statement is true, is its converse necessarily false? The answer is "no," and examples are not hard to find. "If today is Friday, then tomorrow is Saturday" is true, and so is its converse, "If tomorrow is Saturday, then today is Friday."

If an "if . . . then . . ." statement is true, must its converse be true? Again the answer is "no;" can you think of an example to illustrate this?

Some of the theorems of geometry that we have studied have converses that are provable as theorems, also. For instance, Theorem 2.1, "If two sides of a triangle are equal, then the angles opposite those sides are equal," has a true converse which we proved as Theorem 2.4: "If a triangle has two equal angles, then the two sides opposite those angles are equal."

Another alteration of an "if . . . then . . ." statement is of interest to us. If we exchange the antecedent and the consequent, and negate them both (that is, incorporate a "not" into each of them) we form another "if . . . then . . ." statement, called the *contrapositive* of the original statement. For example, the contrapositive of

if it is raining, then it is cloudy, is
if it is not cloudy, then it is not raining.

A statement and its contrapositive are *logically equivalent*; that is, if one is true, the other is also. It is inconceivable that an "if . . . then . . ." statement be true, and yet have a false contrapositive. The two must agree; either both are true, or both are false.

We can prove that the truth of a statement guarantees the truth of the contrapositive by using an indirect argument. Let us assume that $p \to q$ is true, and prove that $\sim q \to \sim p$ is true.

Given: $p \to q$, $\sim q$, both true.

Prove: $\sim p$ is true.

Proof:

STATEMENTS	REASONS
1. p is true.	1. Assumption.
2. $p \to q$.	2. Given.
3. q is true.	3. Modus ponens.
4. $\sim q$ is true.	4. Given.
5. It is false that p is true.	5. It leads to a contradiction (statements 3 and 4).
6. p is false.	6. Only remaining possibility.
7. $\sim p$ is true.	7. The negation of a false statement is true.

The statement and its contrapositive, being logically equivalent, communicate precisely the same information. If we prove a certain "if . . . then . . ." theorem, it is unnecessary to prove the contrapositive as well, since the contrapositive is proved simply by proving the original theorem.

However, some of the proofs in this text are proofs of the contrapositives of theorems proved elsewhere. For example, we proved Theorem 3.18, but this is just the contrapositive of Theorem 3.5. Similarly, Theorem 3.19 is the contrapositive of Theorem 3.6.

A study of the proofs of Theorems 3.18 and 3.19 indicates a procedure similar in form to the proof we have just given. Indeed, proving both a theorem and its contrapositive is just duplicated effort; either proof alone would suffice. The second proofs were included in this text to illustrate the indirect method of proof.

Another rearrangement of the original "if . . . then . . ." statement is of interest to us. If the antecedant and consequent are each negated, the *inverse* of the original statement is formed. The four forms we have studied in this chapter are summarized below.

$p \rightarrow q$	Statement
$q \rightarrow p$	Converse
$\sim p \rightarrow \sim q$	Inverse
$\sim q \rightarrow \sim p$	Contrapositive.

Notice that the inverse of a statement is the contrapositive of the converse; the converse and the inverse are equivalent to each other, just as the statement and its contrapositive are equivalent to each other.

EXERCISES

1. Form the contrapositive of
   (a) If girls are beautiful, then they are not intelligent.
   (b) If it is not cloudy, then it is not raining.
   (c) If you can keep your head while others around you are losing theirs, then you don't understand the situation.
2. Use your imagination to create a true "if . . . then . . ." statement that has a false converse.
3. What is the contrapositive of the converse of "If it is sunny, then we'll go on a picnic?"
4. Prove that the truth of the contrapositive of a statement guarantees the truth of the original statement.
5. Prove that modus tollens is just modus ponens applied to the contrapositive of the "if . . . then . . ." statement.
6. What is the contrapositive of the inverse of an "if . . . then . . ." statement?
7. What is the converse of the inverse of an "if . . . then . . ." statement?

There is another way of looking at these various forms of logical argument. Consider our example, "If I have children, then I am a parent." We may re-write this sentence without changing its meaning: "All people who have children are parents." In this form, our example becomes a statement concerning a relationship between two collections of people—people who have children, and people who are parents. The relationship asserted is that everyone in the first collection belongs to the second collection; that is, the first group is a *subset* of the second. This relationship of inclusion could be diagramed by use of *Euler's circles,* (Fig. 10.2) a device attributed to the eighteenth century Swiss mathematician Leonhard Euler.

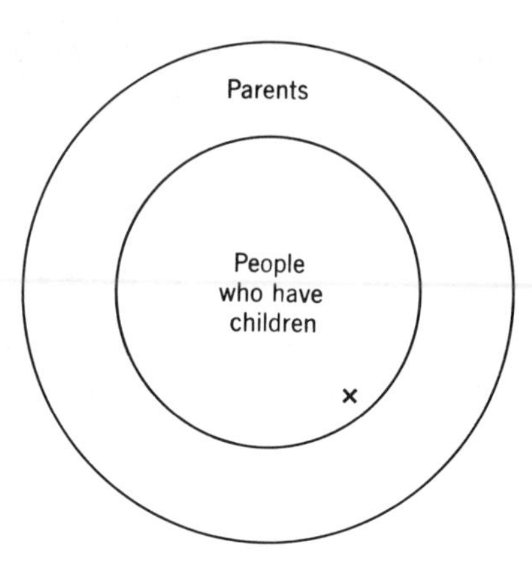

Fig. 10.2

Suppose that another statement is made—"Herman has a child." We could indicate this in our diagram as well, by using a little "×" inside the circle representing people with children. This "×" symbolizes Herman, and its position indicates the meaning of the sentence "Herman has a child." Obviously, since "×" represents an element of the smaller collection, it must therefore be a member of the larger collection, the set of parents. Thus we arrive at our desired conclusion, "Herman is a parent."

We now consider the same set relation as before, "All people who have children are parents," and a second premise, "George is a parent." Let us try and construct Euler circles for this situation. We have a problem deciding where to place the "×" representing George (Fig. 10.3):

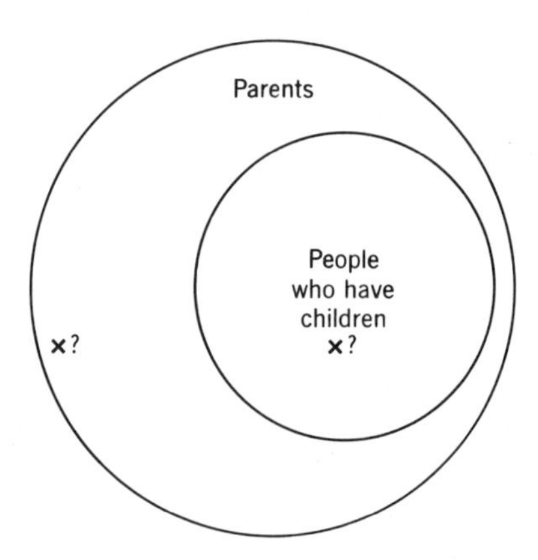

Fig. 10.3

Either position in Fig. 10.3 conveys the meaning of "George is a parent," for either position lies within the boundary of "parents." May we conclude that "George has children?" No, for the "×" need not be placed within the smaller area. May we conclude that "George has no children?" No, for the "×" need not be outside of the area, either. Actually, no conclusion is possible.

Consider the following two premises. "All elephants are friendly" and "All friendly animals are cuddly." An Euler diagram for these statements takes the form of three nested sets (Fig. 10.4):

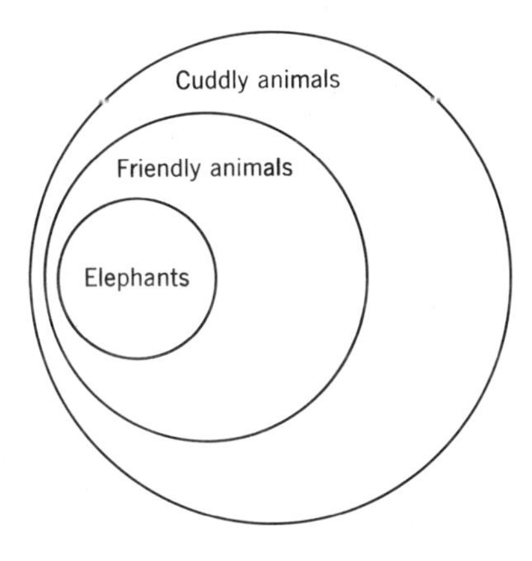

Fig. 10.4

By the relation of these circles, we find our conclusion to be "All elephants are cuddly." (You may disagree with this conclusion, but remember the distinction that we have made between truth and validity.)

Another example:

"All hippopotami are expensive"
"No turtles are expensive."
"My turtle is named Mortimer."

The first sentence diagrams easily (Fig. 10.5).

The second premise says, in effect, that the collection of turtles and the collection of expensive objects are completely separate; there is no overlap at all. This can be diagramed as a circle completely separate from the others, with an "×" to represent Mortimer inside (Fig. 10.6).

The conclusion to be drawn here is, of course, that "Mortimer is not a hippopotamus," which is a great relief to Mortimer.

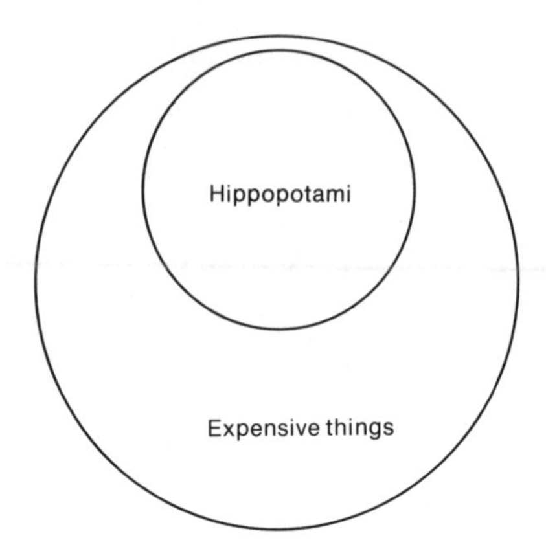

Fig. 10.5

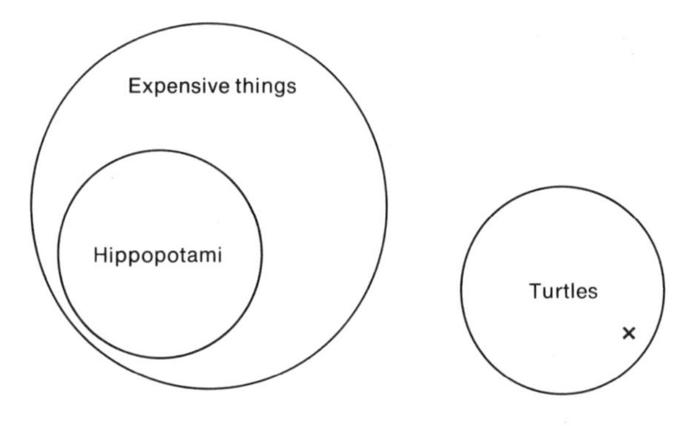

Fig. 10.6

EXERCISES

Draw valid conclusions to the following sets of premises where possible. If no conclusion is possible, indicate *No Conclusion.*

1. If I study hard, then I will pass geometry.
   I do study hard.
   ____

2. If I study hard, then I will pass geometry.
   I will pass geometry.
   ____

3. If I water my garden regularly, the flowers will grow.
   I forgot to plant the seeds.
   ____

4. If I water my garden regularly, the flowers will grow.
   I never water my garden.
   ____

5. If today is Tuesday, I have to go to my piano lesson.
   If I have to go to my piano lesson, I won't be able to study geometry.
   If I don't study geometry, I won't get an A.
   Today is Tuesday.
   ____

6. If he comes to a red light, he stops his car. He did not come to a red light.
   ____

7. No successful students are lazy. All my friends are lazy.
   ____

8. No students who succeed waste their time. No one who tries hard can fail to succeed.
   ____

9. All wholesome fruit is ripe. All these apples are wholesome. No fruit is ripe if it is grown in the shade.
   ____

10. All foreign cars are good. All foreign cars are expensive. Good cars are expensive.
    ____

11. Mathematics teachers are dull. Some dull people are rich. Mathematics teachers are rich.
    ____

The following were inspired by Lewis Carroll (C. L. Dodgson), a 19th-century mathematics teacher, best known as the author of *Alice in Wonderland* and *Through the Looking-Glass*.

12. All my sisters have colds; No one can sing who has a cold.
    ____

13. No bald creature needs a hairbrush; No lizards have hair.
    ____

14. Babies are illogical; Nobody is despised who can manage a crocodile; Illogical persons are despised.
    ____

15. All unripe fruit is unwholesome. All these apples are wholesome. No fruit grown in the shade is ripe.

---

16. Promise breakers are not trustworthy. Wine drinkers are very communicative. All men who keep their promises are honest. No teetotalers are pawn brokers. One can always trust a very communicative person.

Review Test

Classify each of the following as either true or false.

1. The contrapositive of a true statement is necessarily true.
2. The contrapositive of any statement is necessarily true.
3. The converse of a true statement is necessarily false.
4. The contrapositive of the contrapositive of a statement is the original statement.
5. The converse of the converse of a statement is the original statement.
6. If the hypothesis of an argument is true, and the reasoning is valid, then the conclusion must be true also.
7. If the conclusion of a valid argument is false, the hypothesis must be false.
8. If the conclusion of a valid argument is true, the hypothesis must be true.
9. If the hypothesis of a valid argument is false, then the conclusion must certainly be false also.
10. The converse of $r \rightarrow s$ is $s \rightarrow r$.
11. "All unicorns are friendly" is equivalent to "No unicorns are unfriendly."
12. "No unicorns are talkative" is equivalent to "All unicorns are untalkative."
13. If all roses are red, and all violets are blue, then no roses are violets.
14. If no anteaters are redheads, then no redheads are anteaters.
15. If no fish can fly, then no fliers can fish.

11

Coordinate Geometry

If we walk into an elevator, and the operator asks us: "What floor, please?" we only need to supply *one* piece of information to specify our destination completely. "Seventh floor" is a complete instruction; "Seventh floor, and the other end of the building" is beyond the capabilities of a strictly up-and-down elevator.

Locating a destination on a flat surface is somewhat more involved; *two* pieces of information are required. "Main Street, please," is not sufficient information for a taxicab driver; he will certainly ask: "Where on Main Street?"

The city of Chicago is laid out in a manner that makes a location easy to specify. There are two streets that serve as base lines for the city's addresses—Madison, which runs east and west, and State Street, which runs north and south. A position in the city may be specified by noting the distance from these two base lines. "3500 west and 5000 north" specifies a position in the 3500 block west of State Street, and the 5000 block north of Madison.

A similar system is used in geometry to locate a point in a plane. Two base lines, called *axes*, are drawn. One of them, often called the *x-axis*, is usually drawn horizontally, and the other, often called the *y-axis*, is drawn perpendicular to the first. The point of intersection is called the *origin*.

Starting at the origin, an arbitrary unit distance is marked off along the axes, in all four directions. Points are marked as well for twice the unit distance, three times the unit distance, and so on. Using terms like "east" and "west", "north" and "south" to indicate direction would prove inconvenient; instead mathematicians have exploited the nature of signed numbers and have agreed that directions upward or to the right will be regarded as positive, while directions downward or to the left will be negative. Any point can be located by two signed numbers, one indicating its position, left-or-right, and another signed number indicating its position, up-or-down. For example, the point 3 units to the right, and 4 units down, would have a left-right location of +3, and an up-down position of −4. The pair of numbers (+3, −4) is called the pair of *coordinates* of the point, and serves as its unique "address." The point itself is said to be the *graph* of the coordinates. The process of locating and marking the point is called *plotting* the point. The point with coordinate (+3, −4) is plotted in Fig. 11.1, along with points whose coordinates are

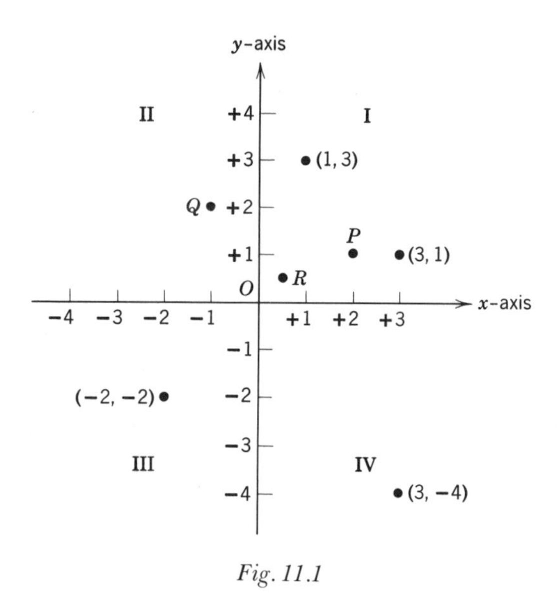

Fig. 11.1

(1, 3), (3, 1), and (−2, −2). What are the coordinates of the origin? What are the coordinates of point P? . . . of point Q? . . . of point R? Notice that (3, 1) and (1, 3) represent two distinct points; the order of the numbers is important. In the pair (x, y) it is understood that x, called the *abscissa*, represents the position relative to the x-axis, and y, called the *ordinate*, represents the position relative to the y-axis. Thus, in Fig. 11.1, point P has coordinates (2, 1), Q is (−1, 2), and R is $(\frac{1}{2}, \frac{1}{2})$.

The two axes divide the plane into four sections, called *quadrants*. These quadrants are numbered in a counterclockwise manner as indicated in Fig. 11.1.

In geometry a line segment is the portion of a line between two points and including the two points. If these two points happened to be points A and B, the line segment was represented with the symbol AB or the symbol BA. We are now going to make a distinction between these symbols. From now on AB will symbolize a line segment starting at point A and going to point B. Line

Fig. 11.2 René Descartes (1596–1650). René Descartes was a French mathematician, theologian, and philosopher, who is credited with applying algebraic methods to geometry. This branch of mathematics is now known as analytic geometry. Descartes died of pneumonia as a result of regular tutoring sessions with Queen Christina of Sweden in a cold room at 5 o'clock in the morning. (New York Public Library Picture Collection).

segment AB has now been assigned a direction (see Fig. 11.3). Line segment BA is the segment starting at point B and going to point A (see Fig. 11.4).

In coordinate geometry a line segment has both a magnitude and direction, just as numbers in algebra. In algebra, plus and minus signs were used to show direction, and they also serve this purpose in coordinate geometry.

If a line is horizontal, it is parallel to the x-axis, and therefore each point

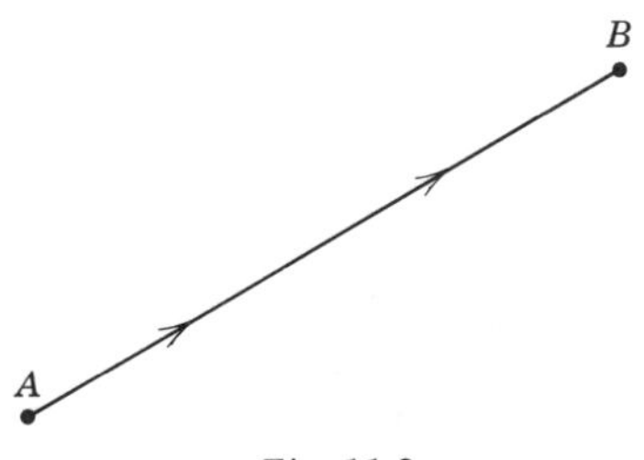

Fig. 11.3

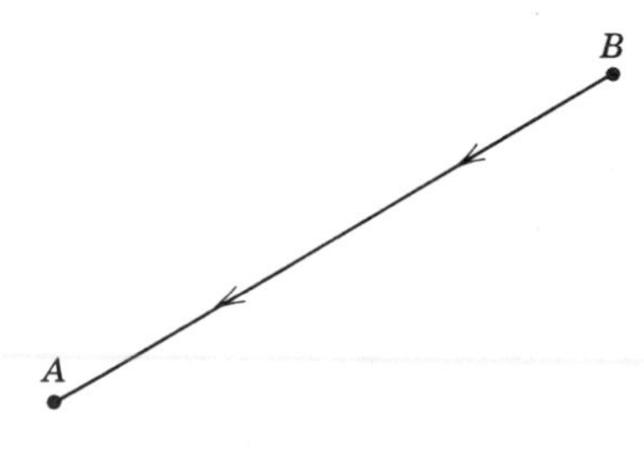

Fig. 11.4

on the line is equidistant from the x-axis. Since the distance of a point P from the x-axis is the y-coordinate of P, it is obvious that a line is horizontal if, and only if, the ordinates of all points on the line are equal. Similarly, a line is vertical if, and only if, the abscissas of all points on the line are equal. Notice in Fig. 11.5 that abscissas of vertical line, v, are equal, and the ordinates of horizontal line, h, are equal.

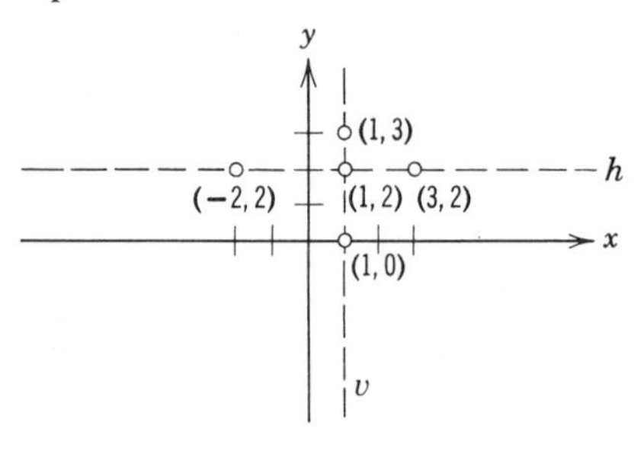

Fig. 11.5

Definition 11.1 The *directed distance of a horizontal line segment from one point to a second* is the x-coordinate of the second minus the x-coordinate of the first.

The directed distance of the line segment AB in Fig. 11.6 is the x-coordinate of B minus the x-coordinate of A. $(6 - 1 = 5)$

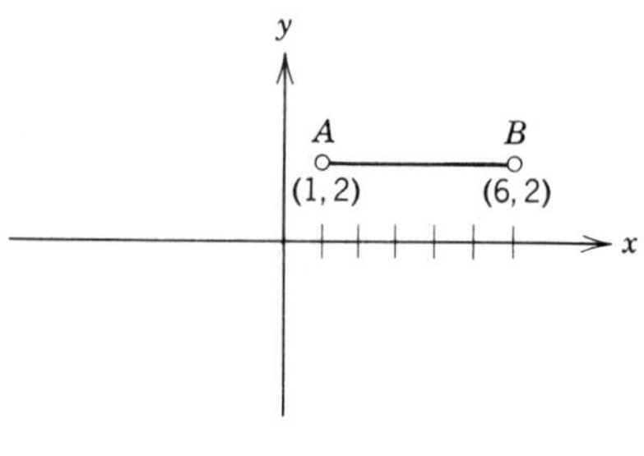

Fig. 11.6

Definition 11.2 The *directed distance of a vertical line segment from one point to a second* is the y-coordinate of the second minus the y-coordinate of the first.

The directed distance of the line segment AB in Fig. 11.7 is the y-coordinate of B minus the y-coordinate of A. $[-4-(+5)] = -4-5 = -9$

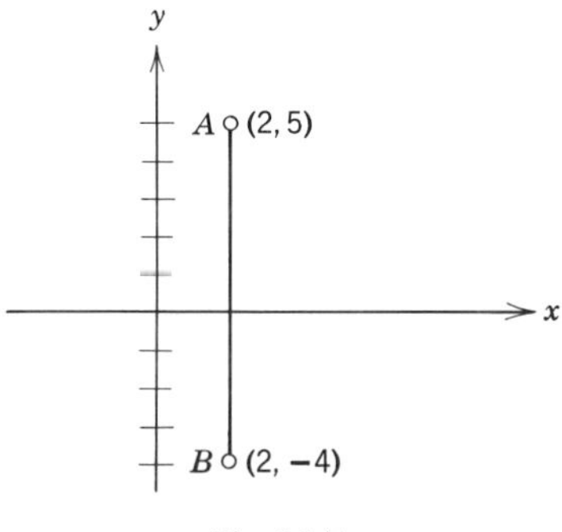

Fig. 11.7

Sometimes it is interesting to look at a picture of an equation. This is possible in mathematics with the use of a graph. Suppose that we wish to graph the equation $x-y=-4$. We begin by finding several number pairs that satisfy the equation. For example, if $x=3$, then $y=7$. If $x=0$, then $y=4$. If $x=-3$, then $y=1$. It is clear that there is no limit to the number pairs that can be found to satisfy the equation. These number pairs can be shown conveniently in a table such as the one in Fig. 11.8. After we have found several number

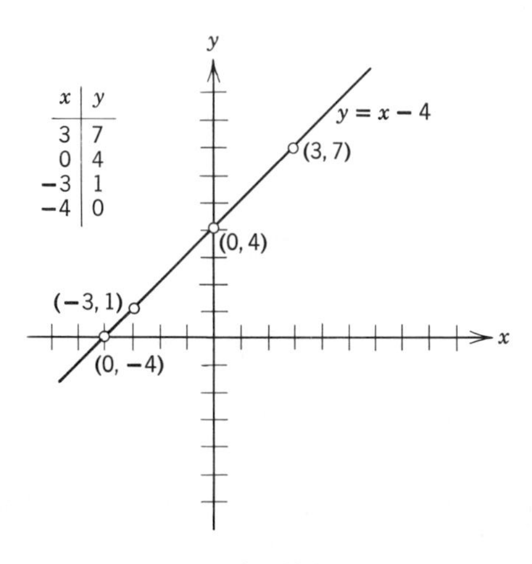

Fig. 11.8

pairs for our table, we plot each individual pair. After plotting several of them it becomes apparent that they all lie in a straight line. This line is called the graph of the equation. The *graph of an equation* is a line or curve such that (1) all points that satisfy the equation lie on the line or curve and (2) all points that lie on the line or curve satisfy the equation.

To graph the equation $3x - y = -1$, we first construct a table of values similar to the one at the right. We then plot each point and draw the line joining these points.

x	y
0	1
2	7
-1	-2
-3	-8

The construction of the graph is left to the reader.

An equation of the form $Ax + By + C = 0$, where A and B are numbers not both zero, is called a *linear equation*. We prove at the end of this chapter that all linear equations must have straight lines for their graphs.

To find the midpoint of a horizontal line segment, we add the abscissas together and then divide by two. Because the line segment is horizontal, the ordinate of the midpoint is the same as the ordinate of the endpoints of the

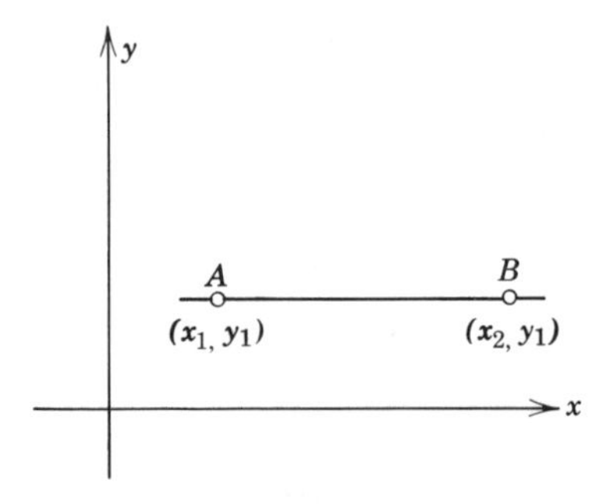

Fig. 11.9

line segment. The midpoint of the line segment in Fig. 11.9 has the coordinates:

$$\left(\frac{x_1 + x_2}{2}, y_1\right)$$

To find the midpoint of a vertical line segment we find the average ordinate. Because the line segment is vertical, the abscissa of the midpoint is the same as

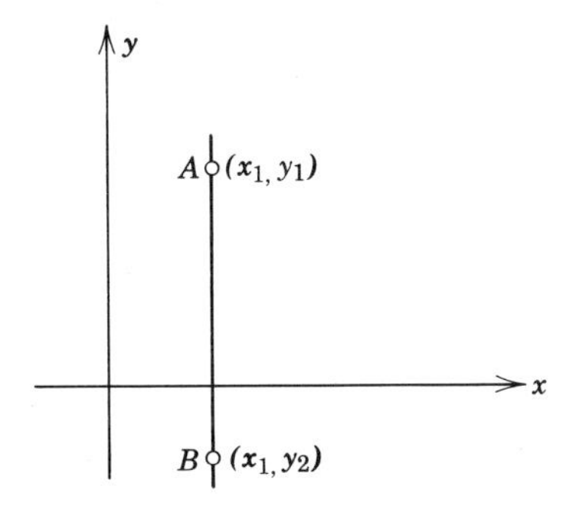

Fig. 11.10

the abscissa of the endpoints of the line segment. The midpoint of the line segment in Fig. 11.10 has the coordinates:

$$\left(x_1, \frac{y_1 + y_2}{2}\right)$$

We can prove the above statements by finding the distance from one end of the line segment to the suspected midpoint and by showing that this is equal to the distance from the suspected midpoint to the other end of the line segment.

For example, a proof that the midpoint of a line segment $P(x_1, y_1)$ $Q(x_2, y_1)$ has the coordinates

$$\left(\frac{x_1 + x_2}{2}, y_1\right)$$

may be written as follows (see Fig. 11.11):

1. By definition the distance from C to Q is:

$$x_2 - \frac{x_1 + x_2}{2} = \tfrac{1}{2}x_2 - \tfrac{1}{2}x_1$$

2. By definition the distance from P to C is:

$$\frac{x_1 + x_2}{2} - x_1 = \tfrac{1}{2}x_2 - \tfrac{1}{2}x_1$$

3. Because these distances are equal, C is the midpoint of the line segment AB.

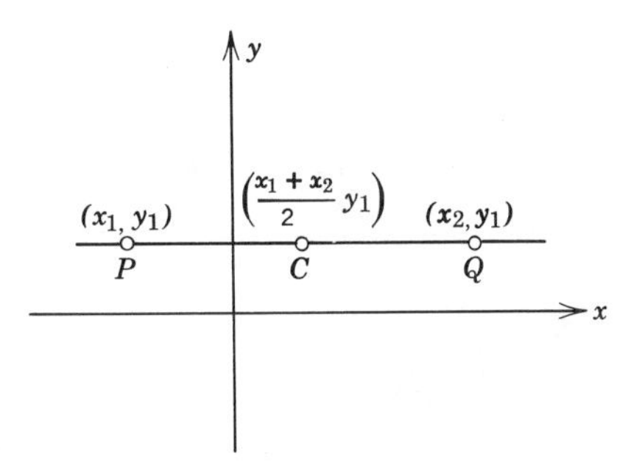

Fig. 11.11

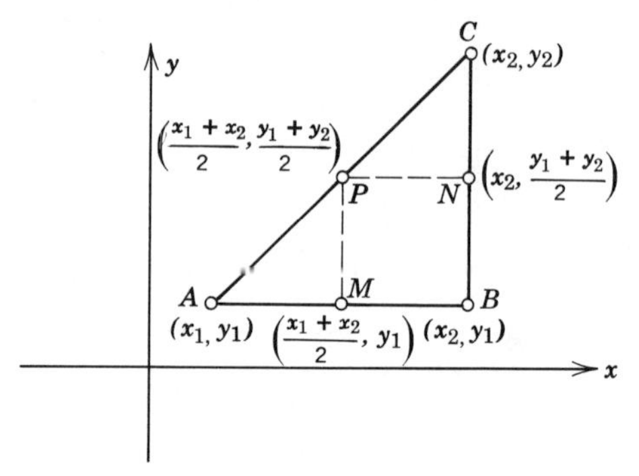

Fig. 11.12

From Fig. 11.12 we can show that the midpoint of a line segment that is not horizontal or vertical has the same abscissa as the midpoint of the horizontal line segment joining point A with point B and the same ordinate as the midpoint of the vertical line segment joining point B with point C. Therefore, the midpoint of the line segment joining A with C has the coordinates:

$$x = \frac{x_1 + x_2}{2} \quad y = \frac{y_1 + y_2}{2}$$

Proof:

STATEMENTS	REASONS
1. $CN = PM$.	1. Why?
2. $PN = AM$.	2. Why?
3. $\angle AMP = \angle PNC$.	3. Why?
4. $\triangle AMP \cong \triangle PNC$.	4. Why?
5. $AP = PC$.	5. Why?
6. P is the midpoint of segment AC.	6. Why?

EXERCISES

Locate (plot) the following points:

1. (0, 3).
2. (−1, −1).
3. (−1, 2).
4. (2, 1).
5. (0, 1).
6. (2, 0).
7. (−3, 0).
8. (1, 0).

9. $(1, -1)$.
10. $(-1, 0)$.
11. $(-2, 1)$.
12. $(3, 0)$.
13. $(1, 2)$.
14. $(-2, 0)$.

If these points are plotted correctly, the dots representing these points will form the shape of a Christmas tree.
Graph the equations

15. $y = x + 1$.
16. $y = 5x - 4$.
17. $y = 3x + 3$.
18. $y = \frac{1}{2}x - 1$.
19. $y = x$.
20. $2y = 3x - 1$.

Find the midpoint of the line segments with the following endpoints:

21. $(6, 4)$ and $(5, 8)$.
22. $(-3, 6)$ and $(6, -3)$.
23. $(-7, 9)$ and $(7, 9)$.
24. $(5, -3)$ and $(-4, -4)$.
25. $(-5, -2)$ and $(-11, -23)$.
26. $(33, 78)$ and $(-21, -86)$.

We now introduce a concept known as the slope of a line.

Definition 11.3 The *slope*, m, of a nonvertical line segment determined by points $P\ (x_1, y_1)$ and $Q(x_2, y_2)$ is given by

$$m = \frac{y_2 - y_1}{x_2 - x_1}$$

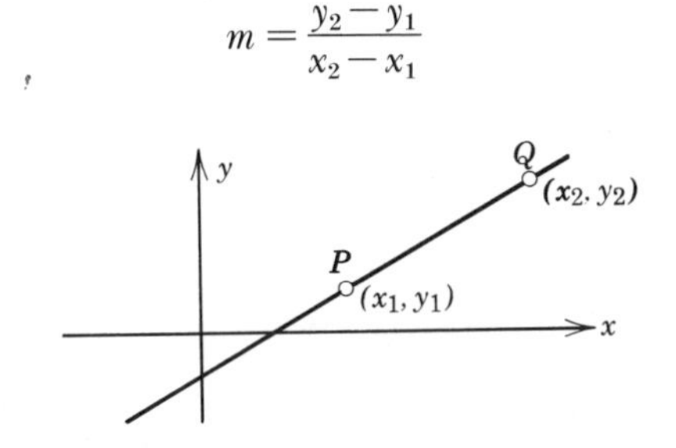

Notice that the slope of a line is the ratio of the difference of the ordinates to the difference of the abscissas of two points on the line, *taken in the same order*.

Since the abscissas are all equal on a vertical line, the difference of the abscissas is zero. Because a fraction with a denominator of zero is meaningless, a vertical line has no slope.

Axiom 11.1 On a nonvertical line, all segments of the line have the same slope.

With the definition of slope and the above axiom, it is possible to write the equation of a line that passes through two known points. Suppose we know that a line passes through the points $P(2, -5)$ and $Q(4, 3)$ (see Fig. 11.13).

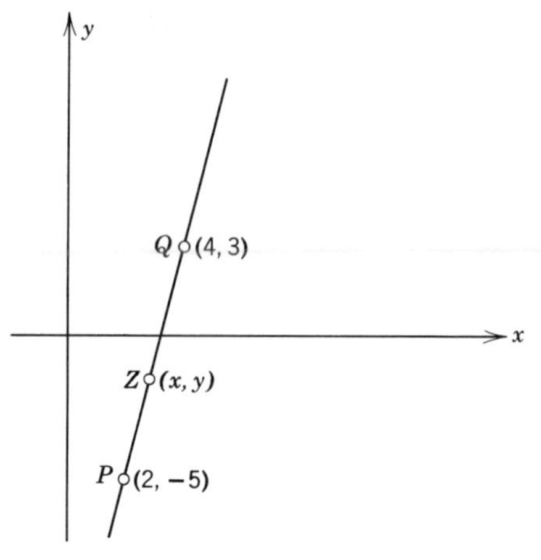

Fig. 11.13

We can pick any third point on the line and call it $Z(x, y)$. First we find the slope of the line segment from P to Q and then find the slope of the line segment from either P or Q to Z. By the above axiom these slopes must be equal. We set the two slopes equal to each other and simplify. The resulting equation is the equation of the line.

$$\text{slope } PQ = \frac{3-(-5)}{4-2} = \frac{4}{1} \qquad \text{slope } QZ = \frac{y-3}{x-4}$$

$$\frac{y-3}{x-4} = \frac{4}{1}$$

$$4x - y - 13 = 0$$

Suppose that we wish to write the equation of a nonvertical line that passes through some point $Q(x_1, y_1)$ and has a slope m. Pick any other point on the line and call it $P(x, y)$.

The definition of slope states that

$$m = \frac{y-y_1}{x-x_1}.$$

After multiplying both sides of the equation by $(x-x_1)$ we obtain what is called the *point slope form* of the straight line:

$$y - y_1 = m(x - x_1)$$

This form allows us to write the equation of the desired line. Suppose we wish to write the equation of a line passing through point $Q(-2, 5)$ and with a slope of 3. After substituting 3 for m and the coordinates of Q for y_1 and x_1, we have

$$y-5=3[x-(-2)]$$
$$y-5=3(x+2)$$
$$3x-y+11=0$$

If the product of two numbers is 1, the numbers are said to be *reciprocals* of each other. If the product of two numbers is -1, the numbers are said to be *negative reciprocals* of each other. It can be shown in advanced mathematics that two lines are perpendicular if their slopes are negative reciprocals of each other. At this time, however, it is best to accept this as an axiom and attempt to show by an example that the above statement is true.

Axiom 11.2 Two nonvertical lines are perpendicular if, and only if, their slopes are negative reciprocals.

Examples: Graph $\begin{cases} 2x+y=2 \\ x-2y=4 \end{cases}$ Observe their graphs, and determine and compare their slopes.

Graph $\begin{cases} 3x-2y=0 \\ 2x+3y=2 \end{cases}$ Observe their graphs, and determine and compare their slopes.

It can also be shown that lines with the same slope never meet and are therefore parallel.

Axiom 11.3 Two nonvertical lines are parallel if, and only if, they have the same slope. Vertical lines are parallel.

Examples: Graph $\begin{cases} 2x-y=2 \\ 4x-2y=0 \end{cases}$ Observe their graphs, and determine and compare their slopes.

Graph $\begin{cases} x+y=0 \\ 3x+3y=3 \end{cases}$ Observe their graphs, and determine and compare their slopes.

Consider any two points in the plane (see Fig. 11.14). Call one point $A(x_2, y_2)$ and call the other point $B(x_1, y_1)$. We can then find a third point that will be

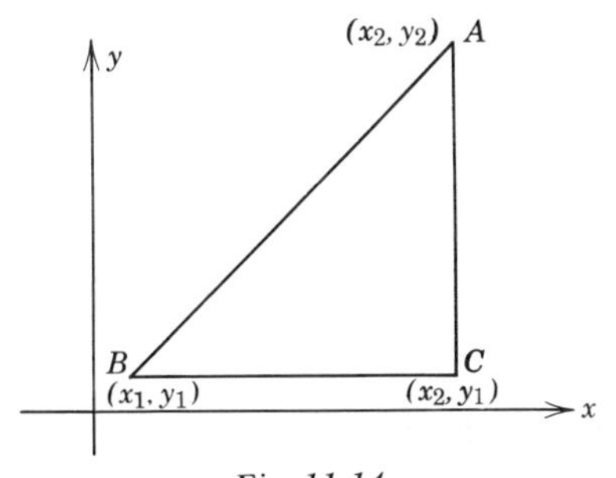

Fig. 11.14

the vertex of a right angle in a right triangle. By inspection, the coordinates of such a point are (x_2, y_1). By definition, the directed distance from point B to point C is $x_2 - x_1$. Also by definition, the directed distance from point C to point A is $y_2 - y_1$. With these facts we can now apply the Pythagorean theorem:

$$\overline{AB}^2 = (x_2 - x_1)^2 + (y_2 - y_1)^2$$

If we call $AB = d$ and take the square root of both sides of the equation, the resulting formula will give us a way to find the *undirected* distance between any two points.

$$d = \sqrt{(x_2 - x_1)^2 + (y_2 - y_1)^2}$$

To find the undirected distance between (3, −2) and (7, 1) we apply the distance formula as follows:

$$\begin{aligned} d &= \sqrt{(7-3)^2 + [1-(-2)]^2} \\ &= \sqrt{16+9} \\ &= 5 \end{aligned}$$

EXERCISES

Find the equations of lines with the following conditions:

1. Passing through (2, 1) and having slope 3.
2. Passing through (−3, 4) and having slope $-\frac{1}{2}$.
3. Passing through (−4, 1) and (6, 3).
4. Passing through (3, 7) and (5, 2).
5. Passing through (2, 1) and having a slope of zero.

Find the undirected distance between the points:

6. (2, 5) and (8, 11).
7. (2, −5) and (13, 12).
8. (3, −4) and (3, 4).
9. (−2, −5) and (−7, 12).
10. (−12, −17) and (0, −8).

There are many geometric theorems that can be proved by using coordinate methods. We now prove some of the easier ones. We have already proved most of them by other methods.

Theorem 3.32

The diagonals of a rectangle are equal.

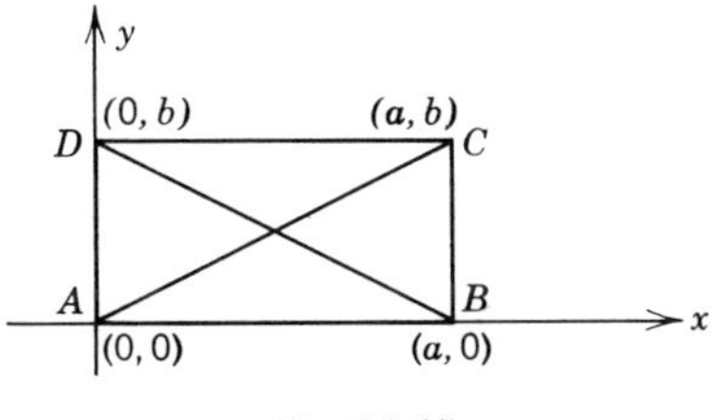

Fig. 11.15

Since $ABCD$ is a rectangle, we can assign coordinates as shown in Fig. 11.15. By the distance formula

$$DB = \sqrt{(0-a)^2+(b-0)^2} = \sqrt{a^2+b^2}$$

$$CA = \sqrt{(0-a)^2+(0-b)^2} = \sqrt{a^2+b^2}$$

Hence $AC = BD$ and the diagonals of a rectangle are equal.

Theorem 2.3

The perpendicular bisector of the base of an isosceles triangle passes through the vertex.

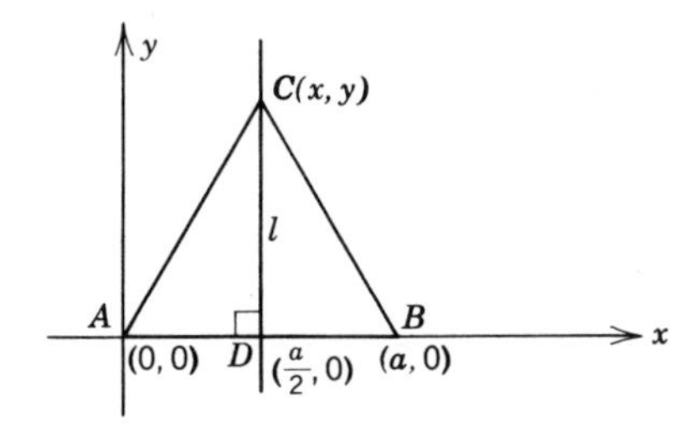

Fig. 11.16

Since $\triangle ABC$ is isosceles it is true that $AC = BC$. If we assign the coordinates (0, 0) to point A, $(a, 0)$ to point B and (x, y) to point C, we have

$$(AC)^2 = (x-0)^2+(y-0)^2 = x^2+y^2$$

$$(BC)^2 = (x-a)^2+(y-0)^2 = a^2-2ax+x^2+y^2$$

Since $AC = BC$, then $(AC)^2 = (BC)^2$ and

$$x^2 + y^2 = a^2 - 2ax + x^2 + y^2$$
$$0 = a^2 - 2ax$$
$$0 = a - 2x$$
$$x = \frac{a}{2}$$

and the coordinates of point C can be written as $(a/2, y)$.

We now write the equation of the perpendicular bisector l, of AB. Since l is a bisector, it must pass through the midpoint of AB so the coordinates of point D are $(a/2, 0)$. The line l is parallel to the y-axis (why?), and therefore all points with abscissas equal to $a/2$ must lie on the line l. Since C has an abscissa of $a/2$, C must lie on the line l.

Theorem 3.23

The opposite sides of a parallelogram are equal.

Since $ABCD$ is a parallelogram, we are justified in using the coordinates in Fig. 11.17: Since $DC \parallel AB$ we are justified in giving points D and C the same

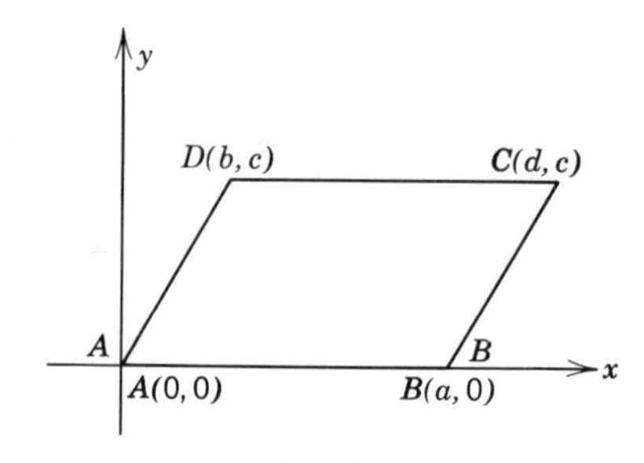

Fig. 11.17

ordinate. Since $AD \parallel BC$, we know that the slope of AD equals the slope of BC. (Why?)

$$\text{slope } AD = \frac{c-0}{b-0} = \frac{c-0}{d-a} = \text{slope } BC$$

and

$$\frac{c}{b} = \frac{c}{d-a}$$

Multiplying both sides by $b(d-a)$, and dividing by c,

$$bc = (d-a)c$$

or

$$b = d - a \tag{1}$$

Since AB and DC are horizontal lines,

$$DC = d - b = d - (d - a) = a$$
$$AB = a - 0 = a$$

Hence

$$AB = DC$$

AD can be shown equal to BC in a similar manner, or we may use the distance formula, and a substitution involving relation (1), above.

Theorem 3.38

A line segment that connects the midpoints of two sides of a triangle is parallel to the third side and half as long.

Arrange the triangle in the coordinate system so that the third side lies on the x-axis, and a vertex is at the origin (Fig. 11.18).

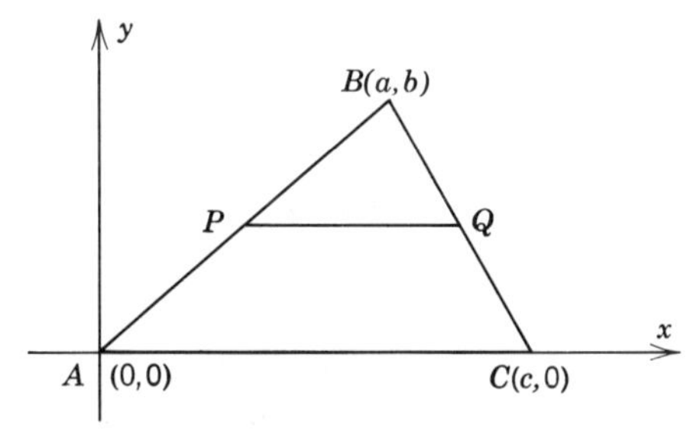

Fig. 11.18

Let PQ be the segment joining the midpoints of the two sides AB and BC.

The abscissa of P is $\frac{a+0}{2}$. Why?

The ordinate of P is $\frac{b+0}{2}$ Why?

The abscissa of Q is $\frac{a+c}{2}$ Why?

The ordinate of Q is $\frac{b+0}{2}$ Why?

Since the ordinates of points P and Q are equal, PQ is parallel to AC. (Why?)

The length of PQ is the difference of the abscissas,

which is $\frac{a+c}{2}-\frac{a}{2}=\frac{c}{2}$ Why?

The length of AC is $c-0=c$ Why?

Therefore $PQ=\frac{1}{2}AC$ Why?

Theorem 3.35

The diagonals of a rhombus are perpendicular to each other.

Since Fig. 11.19 is a rhombus, we know $AB=BC=CD=AD$ (Why?), and we are justified in using the same ordinate for points B and C. (Why?)

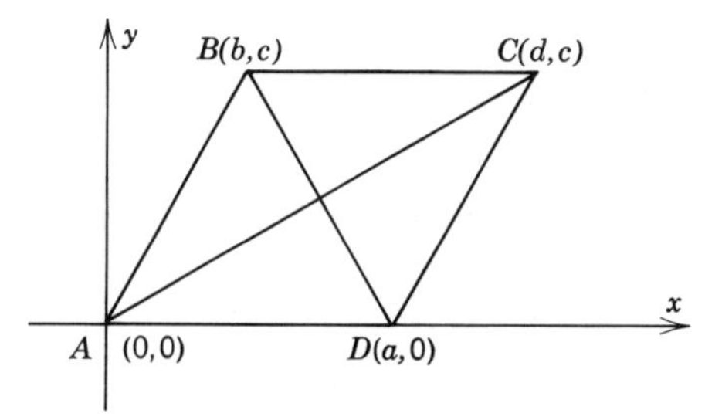

Fig. 11.19

Since $BC=AD$, we know $d-b=a$. (Why?) And, therefore, $d=a+b$. Since $AB=AD$, we know $b^2+c^2=a^2$. (Why?) And, hence, $c^2=a^2-b^2$. (Why?)

The slope of DB is $\frac{c-0}{b-a}=\frac{c}{b-a}$ (Why?)

The slope of AC is $\frac{c-0}{d-0}=\frac{c}{d}$ (Why?)

The product of the slopes is (can you justify each step?)

$$\frac{c}{b-a}\cdot\frac{c}{d}=-\frac{c}{a-b}\cdot\frac{c}{d}=-\frac{c}{a-b}\cdot\frac{c}{a+b}=-\frac{c^2}{a^2-b^2}=-\frac{c^2}{c^2}=-1$$

Since the product of the slopes of the two diagonals is -1, the two diagonals are perpendicular. (Why?)

EXERCISES

Prove the following by coordinate methods.

1. The diagonals of an isosceles trapezoid are equal.

2. The diagonals of a parallelogram bisect each other.
3. The segments joining the midpoints of adjacent sides of any quadrilateral form a parallelogram.
4. The segments joining the midpoints of opposite sides of any quadrilateral bisect each other.
5. A median from the vertex of an isosceles triangle to the base is perpendicular to the base.

Although the following theorem is quite complicated, it is important and warrants our attention. First we need a definition.

Definition 11.4 A *linear equation* is any equation which can be put in the form of $Ax+By+C=0$.

Thus, $2x+3y=7$, $5x=\frac{3}{2}y$, $x=0$, $x+2y+3=0$, are all linear equations.

Theorem 11.1

The locus of points whose coordinates satisfy a given linear equation is a straight line.

Plan: In order to establish the locus we must show that:

1. All number pairs (x, y) that satisfy the equation represent points that lie on the line.
2. All points on the line are represented by number pairs (x, y) that satisfy the equation.

Proof: let (x_1, y_1) and (x_2, y_2) be two different number pairs which satisfy $Ax+By+C=0$. That is,

$$Ax_1+By_1+C=0 \tag{1}$$

and

$$Ax_2+By_2+C=0$$

Subtracting the first equation from the second, we obtain

$$A(x_2-x_1)+B(y_2-y_1)=0 \tag{2}$$

or

$$A(x_2-x_1)=-B(y_2-y_1)$$

or

$$\frac{y_2-y_1}{x_2-x_1}=\frac{-A}{B}$$

provided that the denominators are not zero. If these quotients are meaningful, the left side represents the slope of the line passing through points $A(x_1, y_1)$ and $B(x_2, y_2)$.

We are now ready to attack the locus problem by proving Part 1—that any pair (x, y) which satisfies the equation represents a point on the line AB. Let $P(x', y')$ be an arbitrary point on the graph of $Ax+By+C=0$. Then (x', y') satisfies the equation, and

$$Ax'+By'+C=0$$

Subtracting this from eq. 1, we obtain

$$A(x_1-x')+B(y_1-y')=0$$

or

$$\frac{y_1-y'}{x_1-x'}=-\frac{A}{B}$$

The left side of this equation represents the slope of the line joining $A(x_1, y_1)$ and $P(x', y')$. The right side is identical to the right side of eq. 2. Therefore, we know that the slope of the line AP equals the slope of line AB. Since these have the point A in common, they must be the same line. Since $P(x', y')$ was *any* point whose coordinates satisfy $Ax+By+C=0$, we know that any such point lies on the line joining A and B.

To prove the second half, let $P(x, y)$ be any point on the line AB. We wish to show that x and y satisfy $Ax+By+C=0$. Since we are assuming P to lie on AB, we know by Axiom 11.1 that the slope of PA equals the slope of AB. The slope of AB has been computed as $-A/B$. The slope of PA is $(y_1-y)/(x_1-x)$. Setting these equal, and manipulating,

$$\frac{y_1-y}{x_1-x}=-\frac{A}{B}$$

$$-A(x_1-x)=B(y_1-y)$$

$$Ax-Ax_1=By_1-By$$

$$Ax+By-Ax_1-By_1=0 \tag{3}$$

But, by rearranging eq. 1, we find that

$$C=-Ax_1-By_1$$

Hence, by substitution in eq. 3,

$$Ax+By+C=0$$

We now have shown that any arbitrary point on the line joining A and B has coordinates that satisfy the general linear equation $Ax+By+C=0$. We have proved that the locus of points whose coordinates satisfy a linear equation is a straight line.

Review Test

Classify the following statements as true or false.

1. The ordinate of (3, 7) is 3.
2. The points $(3, -2)$ and $(2, -3)$ lie in the same quadrant.
3. If (x, y) is a point in quadrant I, then x and y are both positive.
4. The origin has no coordinates.
5. None of the points (2, 3), $(-3, -2)$, $(1, -4)$, $(3, -2)$ and $(-5, -3)$ lie in quadrant II.
6. If AB is a directed line segment, then $AB = -BA$.
7. The distance of point (3, 4) from the origin is 5.
8. A line with positive slope rises up and to the right.
9. The slope of the x-axis is zero.
10. The slope of the y-axis is one.
11. The line passing through the origin and point (a, b) is perpendicular to the line passing through (b, a) and (0, 0).
12. The line $y + 2x = 1$ passes through the origin.
13. The line $y + 2x = 1$ and $y + 2x = 3$ are parallel.
14. $y = x^2$ represents a line.
15. $y = k$ represents a vertical line.

Appendix I

Fig. A.1. James A. Garfield (1831–1881). James A. Garfield, was an amateur mathematician as well as the twentieth president of the United States. Garfield's best known contribution to mathematics is his proof of the Pythagorean theorem. (New York Public Library Picture Collection).

The following proof of the Pythagorean theorem has been attributed to James A. Garfield, the twentieth President of the United States.

Given: right triangle ABC.

Prove: $a^2 + b^2 = c^2$.

Construction. Extend side BC to point D so that $BD = CA$. Construct segment of DE so that $DE \perp CD$ and $DE = BC$. Draw segments BE and AB.

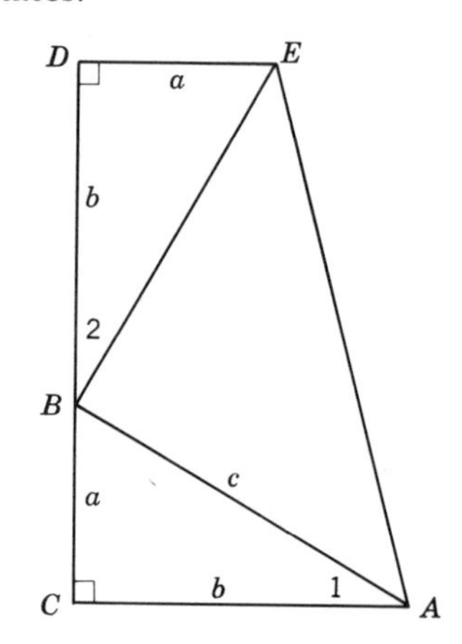

Proof: quadrilateral $ACDE$ is a trapezoid. Why? The area of the trapezoid $ACDE$ is given by

$$A = \tfrac{1}{2}h(b+b') = \tfrac{1}{2}(a+b)(a+b) = \tfrac{1}{2}(a^2+2ab+b^2)$$

The area of the trapezoid can also be found by suming the areas of triangles ACB, BDE, and EBA. After proving $\triangle ABC \cong \triangle BDE$ and observing that $\angle 1 = \angle 2$, $\angle ABE$ can be shown to be a right angle. Why? Since all three of the triangles have a right angle their areas are

$$A(\triangle ACB) = \tfrac{1}{2}ab$$
$$A(\triangle BDE) = \tfrac{1}{2}ab$$
$$A(\triangle BEA) = \tfrac{1}{2}c^2$$

Therefore the area of the trapezoid is

$$A = \tfrac{1}{2}ab + \tfrac{1}{2}ab + \tfrac{1}{2}c^2 = ab + \tfrac{1}{2}c^2$$

By setting the two expressions for the area equal to each other, we have

$$\tfrac{1}{2}(a^2+2ab+b^2) = ab + \tfrac{1}{2}c^2$$

or

$$a^2+2ab+b^2 = 2ab+c^2$$

or

$$a^2+b^2 = c^2.$$

Appendix II

A different way of finding the area of a triangle is credited to Heron of Alexandria (Hero), an early Greek mathematician. His formula gives the area of a triangle as a function of the lengths of its sides. Hero's formula is

$$A = \sqrt{s(s-a)(s-b)(s-c)}$$

where A is the area, s is one-half the perimeter, and a, b, and c are the lengths of the sides of the triangle.

Example: Find the area of a triangle with sides of length

4 inches, 5 inches, and 7 inches.

Solution: One-half the perimeter is

$$s = \tfrac{1}{2}(4+5+7) = 8$$

and therefore the area is

$$A = \sqrt{8(8-4)(8-5)(8-7)} = \sqrt{8(4)(3)(1)} = \sqrt{96} = 4\sqrt{6}. \text{ sq. inches.}$$

Answers for Selected Exercises

CHAPTER I

page 6 **4.** 15 inches **5.** 45 inches **6.** 12 inches **7.** 17.9 inches

pages 14–15 **1.** No **3.** $\angle ABC$ or $\angle B$ **4.** They meet in a point. **6.** $\angle 5$ and $\angle 6$, $\angle 3$ and $\angle 4$, $\angle 3$ and $\angle 6$, etc. **7.** They meet in a point. **8.** They meet in a point. **15.** 6, 3, 6, 3, 5, 7 **16.** 5, 8

page 18 **2.** 150°, 135°, 45° **3.** 60°, 45°, 10° **4.** No, Yes **5.** 30°, 150° **6.** No **7.** No, No **8.** No **9.** 150° **10.** 135° **11.** 67° **12.** 37° **15.** 60°

pages 19–20 **1.** F **2.** T **3.** T **4.** T **5.** F **6.** F **7.** F **8.** F **9.** T **10.** T **11.** F **12.** F **13.** T **14.** F **15.** T **16.** T **17.** T **18.** F **19.** F **20.** F **21.** T **22.** F **23.** T **24.** T **25.** T

CHAPTER II

pages 39–40 **1.** T **2.** F **3.** T **4.** F **5.** F **6.** T **7.** T **8.** F **9.** F **10.** T **11.** F **12.** T **13.** T **14.** F **15.** T **16.** F **17.** F **18.** T **19.** T **20.** F

CHAPTER III

pages 46–47 **1.** $\angle a$ and $\angle h$, $\angle g$ and $\angle b$ **2.** $\angle a$ and $\angle e$, $\angle b$ and $\angle f$, $\angle c$ and $\angle g$, $\angle d$ and $\angle h$ **5.** No **6.** Yes **7.** No **8.** No

pages 50–51 **4.** 30°, 150°

pages 54–55 **5.** 105°, 45°, 105°, 30° **10.** 60°, 60°, 60°; 45°, 45°, 90°

page 58 **1.** 1080° **2.** 1440° **3.** 18 **4.** 360° **5.** 108° **6.** 12 **7.** No **8.** No **9.** Yes **10.** 24° **12.** 6 **13.** 6 **14.** 5

pages 64–65 **7.** 60°, 120°, 60°, 120°

page 70 **6.** 5

page 74 **1.** 14 **4.** 58

pages 76–77 **1.** F **2.** F **3.** T **4.** F **5.** T **6.** T **7.** T **8.** T **9.** F **10.** F **11.** F **12.** T **13.** F **14.** T **15.** F **16.** T **17.** T **18.** F **19.** T **20.** F

CHAPTER IV

pages 84–86 **a.** 15 sq. in. **b.** 6 sq. in. **c.** 12 sq. in. **d.** 23 sq. in. **e.** $52\frac{1}{2}$ sq. ft. **f.** $18+\frac{9}{2}\pi$ sq. mi. **g.** 84 sq. ft. **h.** 35/2 sq. ft. **i.** 54 sq. in. **j.** $70\frac{1}{2}$ sq. in. **k.** If s is the side of the smallest square, the area is $341s^2$. **3.** 144 sq. in. **4.** 49/2 sq. in. **5.** 2 in. **6.** 4, 2, 4, 1

pages 92–93 **2.** $2\sqrt{2}$ in. **3.** 2 in., $2\sqrt{3}$ in. **4.** $2\sqrt{3}$ sq. in. **5.** 6 in. **6.** $\sqrt{3}$ in. **7.** $\sqrt{3}$ sq. in. **10.** $\sqrt{3}$ in. **11.** $(16/3)\sqrt{3}$ in. **12.** 6 ft. **13.** 12 ft. **14.** Approximately 453 ft. plus waste **15.** Yes **16.** $6\sqrt{10}$ in.

page 94 **1.** T **2.** F **3.** T **4.** F **5.** F **6.** F **7.** F **8.** F **9.** T **10.** T **11.** F **12.** F **13.** T **14.** F **15.** T **16.** T **17.** T **18.** T **19.** F **20.** T

CHAPTER V

pages 101–102 **1.** 5 **2.** 3 or −3 **3.** 11/3 **4.** −3 **8.** Yes **9.** Theorems 5.2, 5.3, 5.4 **10.** 0 **11.** Impossible **14.** 5 **15.** 4, −1 **16.** No **17.** 51 **18.** All values of x.

pages 105–106 **3.** 63/4 **4.** $CE = 240/17$, $EB = 100/17$ **5.** Not enough information given **6.** $AD = 40/3$, $DB = 20/3$ **13.** $EF = 20/3$, $GF = 16/3$, $AB = 4$

pages 112–113 **1.** $\sqrt{117}$, $\sqrt{52}$, 6 **2.** 6, 6, $6\sqrt{2}$ **3.** 12, 27 **4.** 9 **5.** 128/3 **6.** Yes **14.** No **18.** 174 ft. **20.** 150 ft.

pages 113–114 **1.** T **2.** T **3.** T **4.** F **5.** F **6.** T **7.** T **8.** F **9.** F **10.** T **11.** T **12.** F **13.** T **14.** T **15.** T **16.** F **17.** F **18.** F **19.** T **20.** T

CHAPTER VI

pages 121–123 **5.** 45°/2, 45° **6.** 70°, 50° **7.** 90° **8.** 20° **9.** 120° **10.** 60° **11.** 110° **12.** 130° **13.** 50°, 30° **14.** 30°, 15° **15.** 40°

page 128 **5.** 6 **6.** 8 **7.** 1 **8.** 220° **9.** 150°

pages 131–132 **1.** 2 **2.** Infinite **3.** 0, 1 or 2 **4.** 1 or 2 **7.** 20° **9.** 100°, 50° **10.** 140° **11.** Yes

pages 136–137 **2.** 4 in. **4.** $4\sqrt{3}$ **5.** $4\sqrt{10}$ **6.** $78\frac{3}{4}$ **7.** 3 **8.** 4 **9.** 150°

page 138 **1.** T **2.** F **3.** T **4.** F **5.** F **6.** T **7.** F **8.** F **9.** T **10.** F **11.** F **12.** F **13.** F **14.** T **15.** T **16.** T **17.** T **18.** F **19.** T **20.** F

CHAPTER VII

page 145 **1.** $x < 4$ **2.** $x < -10/3$ **3.** $x < 10$ **4.** $x < \frac{2}{5}$ **5.** $x > 30/19$ **6.** $x > 10/3$ **7.** $x > 0$

pages 153–154 **1.** F **2.** T **3.** F **4.** T **5.** F **6.** T **7.** T **8.** T **9.** T **10.** T **11.** F **12.** F **13.** T **14.** T **15.** T **16.** T **17.** F **18.** T **19.** T **20.** F

CHAPTER VIII

page 159 **4.** A line parallel to the given lines and midway between them.

page 164 **8.** No **11.** Yes **12.** $\frac{8}{3}$ inches . $\frac{4}{3}$ inches.

pages 164–165 **1.** F **2.** T **3.** T **4.** F **5.** T **6.** F **7.** T **8.** T **9.** T **10.** T **11.** T **12.** T **13.** F **14.** T **15.** F

CHAPTER IX

page 169 **7.** Yes **10.** No

page 173 **3.** 45°, 36° **5.** $48\sqrt{3}$ sq. in. **6.** 750 sq. in. **8.** 2500π sq. yds. **9.** $245/36\pi$ sq. in. **10.** $r = 2$

page 174–175 **1.** F **2.** T **3.** T **4.** T **5.** F **6.** F **7.** T **8.** T **9.** F **10.** T **11.** F **12.** F **13.** T **14.** F **15.** T

CHAPTER X

pages 178–179 **1.** Yes, only Martians know. **2.** No, No **4.** Yes **6.** Tomorrow is a holiday. **7.** No conclusion. **8.** I will pass geometry. **9.** The mail is late.

page 181 **1.** I get a diamond. **2.** It is false that it is hot and humid. **3.** No conclusion. **4.** Geometry is not fun.

pages 186–188 **2.** No conclusion. **3.** No conclusion. **4.** No conclusion. **5.** I won't get an A. **6.** No conclusion. **7.** None of my friends are successful. **10.** No conclusion.

page 188 **1.** T **2.** F **3.** F **4.** T **5.** T **6.** T **7.** T **8.** F **9.** F **10.** T **11.** T **12.** T **13.** T **14.** T **15.** F

CHAPTER XI

pages 196–197 **21.** $(\frac{11}{2}, 6)$ **22.** $(\frac{3}{2}, \frac{3}{2})$ **23.** $(0, 9)$ **24.** $(\frac{1}{2}, -\frac{7}{2})$ **25.** $(-8, -\frac{25}{2})$ **26.** $(6, 6)$

page 200 **1.** $3x - y = 5$ **2.** $x + 2y = 5$ **3.** $x - 5y = -9$ **5.** $y = 1$ **6.** $6\sqrt{2}$ **7.** $\sqrt{410}$ **8.** 8 **10.** 15

page 207 **1.** F **2.** T **3.** T **4.** F **5.** T **6.** T **7.** T **8.** T **9.** T **10.** F **11.** F **12.** F **13.** T **14.** F **15.** F

Index